Georg Eder

From building blocks to 2D networks

Georg Eder

From building blocks to 2D networks
An STM study on the interactions at the nanoscale

Südwestdeutscher Verlag für Hochschulschriften

Impressum / Imprint
Bibliografische Information der Deutschen Nationalbibliothek: Die Deutsche Nationalbibliothek verzeichnet diese Publikation in der Deutschen Nationalbibliografie; detaillierte bibliografische Daten sind im Internet über http://dnb.d-nb.de abrufbar.
Alle in diesem Buch genannten Marken und Produktnamen unterliegen warenzeichen-, marken- oder patentrechtlichem Schutz bzw. sind Warenzeichen oder eingetragene Warenzeichen der jeweiligen Inhaber. Die Wiedergabe von Marken, Produktnamen, Gebrauchsnamen, Handelsnamen, Warenbezeichnungen u.s.w. in diesem Werk berechtigt auch ohne besondere Kennzeichnung nicht zu der Annahme, dass solche Namen im Sinne der Warenzeichen- und Markenschutzgesetzgebung als frei zu betrachten wären und daher von jedermann benutzt werden dürften.

Bibliographic information published by the Deutsche Nationalbibliothek: The Deutsche Nationalbibliothek lists this publication in the Deutsche Nationalbibliografie; detailed bibliographic data are available in the Internet at http://dnb.d-nb.de.
Any brand names and product names mentioned in this book are subject to trademark, brand or patent protection and are trademarks or registered trademarks of their respective holders. The use of brand names, product names, common names, trade names, product descriptions etc. even without a particular marking in this works is in no way to be construed to mean that such names may be regarded as unrestricted in respect of trademark and brand protection legislation and could thus be used by anyone.

Coverbild / Cover image: www.ingimage.com

Verlag / Publisher:
Südwestdeutscher Verlag für Hochschulschriften
ist ein Imprint der / is a trademark of
AV Akademikerverlag GmbH & Co. KG
Heinrich-Böcking-Str. 6-8, 66121 Saarbrücken, Deutschland / Germany
Email: info@svh-verlag.de

Herstellung: siehe letzte Seite /
Printed at: see last page
ISBN: 978-3-8381-3727-8

Zugl. / Approved by: München, LMU, Diss., 2013

Copyright © 2013 AV Akademikerverlag GmbH & Co. KG
Alle Rechte vorbehalten. / All rights reserved. Saarbrücken 2013

Abstract

The aim of this work is to further the understanding of the important parameters in the formation process of 2D nanostructures and therewith pioneer for novel applications. Such 2D nanostructures can be composed of specially designed organic molecules, which are adsorbed on various surfaces. In order to study true 2D structures, monolayers were deposited. Their properties have been investigated by scanning tunneling microscopy (STM) under ultra-high vacuum (UHV) conditions as well as under ambient conditions. The latter is a highly dynamic environment, where several parameters come into play. Complementary surface analysis techniques such as low-energy electron diffraction (LEED), X-Ray photo-emission spectroscopy (XPS), and Raman spectroscopy were used when necessary to characterize these novel molecular networks.

In order to conduct this type of experiments, high technical requirements have to be fulfilled, in particular for UHV experiments. Thus, the focus is on a drift-stable STM, which lays the foundation for high resolution STM topographs. Under ambient conditions, the liquid-solid STM can be easily upgraded by an injection add-on due to the highly flexible design. This special extension allows for adding extra solvent without impairing the high resolution of the STM data. Besides the device, also the quality of the tip is of pivotal importance. In order to meet the high requirements for STM tips, an *in vacuo* ion-sputtering and electron-beam annealing device was realized for the post-preparation of scanning probes within one device. This two-step cleaning process consists of an ion-sputtering step and subsequent thermal annealing of the probe.

One study using this STM setup concerned the incorporation dynamics of coronene (COR) guest molecules into pre-existent pores of a rigid 2D supramolecular host networks of trimesic acid (TMA) as well as the larger analogous benzenetribenzoic acid (BTB) at the liquid-solid interface. By means of the injection add-on the additional solution containing the guest molecules was applied to the surface. At the same time the incorporation process was monitored by the STM. The incorporation dynamics into geometrically perfectly matched pores of trimesic acid as well as into the substantially larger pores of benzentribenzoic acid exhibit a clearly different behavior. For the BTB network instantaneous incorporation within the temporal resolution of the experiment was observed; for the TMA network, however, intermediate adsorption states of COR could be visualized before the final adsorption state was reached.

A further issue addressed in this work is the generation of metal-organic frameworks (MOFs) under ultra-high vacuum conditions. A suitable building block therefore is an aromatic trithiol, *i.e.* 1,3,5-tris(4-mercaptophenyl)benzene (TMB). To understand the specific role of the substrate, the surface-mediated reaction has been studied on Cu(111) as well as on Ag(111). Room temperature deposition on both substrates results in densely packed trigonal structures. Yet, heating the Cu(111) with the TMB molecules to moderate temperature (150 °C) yields two different porous metal coordinated networks, depending on the initial surface coverage. For Ag(111) the first structural change occurs after annealing the sample at 300 °C. Here, several disordered structures with partially covalent disulfur bridges were identified.

Proceeding further in the scope of increasing interaction strength between the building blocks, covalent organic frameworks (COFs) were studied under ultra-high vacuum conditions as well as under ambient conditions. For this purpose, a promising strategy is covalent coupling through radical addition reactions of appropriate monomers, *i.e.* halogenated aromatic molecules such as 1,3,5-tris(4-bromophenyl)benzene (TBPB) and 1,3,5-tris(4-iodophenyl)benzene (TIPB). Besides the correct choice of a catalytic surface, the activation energy for the scission of the carbon-halogen bonds is an essential parameter. In the case of ultra-high vacuum experiments, the influence of substrate temperature, material, and crystallographic orientation on the coupling reaction was studied. For reactive Cu(111) and Ag(110) surfaces room temperature deposition of TBPB already leads to a homolysis of the C-Br bond and subsequent formation of proto-polymers. Applying additional heat facilitates the transformation of proto-polymers into 2D covalent networks. In contrast, for Ag(111) just a variety of self-assembled and rather poorly ordered structures composed of intact molecules has emerged. The deposition onto substrates held at 80 K has never resulted in proto-polymers.

For ambient conditions, the polymerization reaction of 1,3,5-tri(4-iodophenyl)benzene (TIPB) on Au(111) was studied by STM after drop-casting the monomer onto the substrate held either at room temperature or at 100 °C. For room temperature deposition only poorly ordered non-covalent arrangements were observed. In accordance with the established UHV protocol for halogenated coupling reaction, a covalent aryl-aryl coupling was accomplished for high temperature deposition. Interestingly, these covalent aggregates were not directly adsorbed on the Au(111) surface, but attached on top of a chemisorbed monolayer comprised of iodine and partially dehalogenated TIPB molecules. For a detailed analysis of the processes, the temperature dependent dehalogenation reaction was monitored by X-ray photoelectron spectroscopy under ultra-high vacuum conditions.

Keywords

Scanning tunneling microscopy, ultra-high vacuum, self-assembly, supramolecular, host-guest networks, metal-organic frameworks, covalent organic frameworks, graphene.

Contents

1	**Motivation**	**1**
2	**Introduction to scanning tunneling microscopy**	**7**
	2.1 Fundamental principle	7
	2.2 Tunneling through molecules – Contrast mechanism and interpretation	14
	2.3 Instrumentation	21
3	**Experimental Details and Materials**	**27**
	3.1 Ultra-high vacuum versus ambient environment	27
	3.2 STM tip preparation	32
	3.3 Substrates, solvents, and molecules	38
	3.4 Surface and sample preparation	44
4	**Interactions at the nanoscale – from building blocks to 2D networks**	**51**
	4.1 Basic principles – processes at the nanoscale	51
	4.2 Interaction between molecules	61
	4.3 Interaction between molecules and surfaces	68
5	**Two-dimensional molecular structures**	**75**
	5.1 Supramolecular host-guest networks at the liquid-solid interface – Incorporation dynamics	75
	5.2 Halogen versus hydrogen bonded 2D networks	83
	5.3 Organic molecules on graphene terminated substrates	87
	5.4 2D Metal-organic frameworks (MOFs) based on thiolate-copper coordination bonds	90

5.5 2D Covalent Organic Frameworks – COFs 96
    5.5.1 2D COFs under ultra-high vacuum conditions 96
    5.5.2 2D COFs under ambient conditions 103

6 Summary and Outlook   111

7 Publications   115

7.1 Surface mediated synthesis of 2D covalent organic frameworks: 1,3,5-tris(4-bromophenyl)benzene on graphite(001), Cu(111), and Ag(110) 116

7.2 Material- and orientation-dependent reactivity for heterogeneously catalyzed carbon-bromine bond homolysis 120

7.3 A combined ion-sputtering and electron-beam annealing device for the *in-vacuo* post preparation of scanning probes 127

7.4 Extended two-dimensional metal-organic frameworks based on thiolate-copper coordination bonds 132

7.5 Incorporation dynamics of molecular guests into two-dimensional supramolecular host networks at the liquid-solid interface . . . 140

7.6 Solution preparation of two dimensional covalently linked networks by polymerization of 1,3,5-tri(4-iodophenyl)benzene on Au(111) 150

References   159

Acknowledgements   179

Abbreviations

AFM	Atomic force microscope
COFs	Covalent organic frameworks
EDX	Energy dispersive X-ray spectroscopy
ESQC	Elastic scattering quantum chemistry
HOMO	Highest occupied molecular orbital
HOPG	Highly ordered pyrolytic graphite
LDOS	Local density of states
LUMO	Lowest unoccupied molecular orbital
LUT	Lookup table
MOFs	Metal-organic framework
OMBE	Organic molecular beam epitaxy
RAHB	Resonance assisted hydrogen bonding
SAM	Self-assembled monolayer
SEM	Scanning electron microscopy
STM	Scanning tunneling microscope
STS	Scanning tunneling spectroscopy
TST	Transition state theory

CHAPTER 1

Motivation

Science and research are preferentially conducted to improve living conditions and solve upcoming problems. The major challenge that needs to be faced is the demographic growth and associated problems like nutrition, environmental issues, energy production, and responsible exposure of resources. To date, science and technology are intrinsically tied to our daily life and are therefore crucial for the overall progress of mankind. They help to preserve stable democracies all over the world by assuring a constantly growing economy and national prosperity.

Scientific research aims at gaining knowledge of the complexity of nature and transferring these results into technical products. Research has pioneered inventions and discoveries throughout all of time. In the last centuries a quantum leap forward has been achieved in health care, energy generation, communication, and computer technology. Recently, the field of nanotechnology, *i.e.* materials and structures with characteristic dimensions or tolerances of less than 100 nm, has attracted great attention. New insights and inventions in this field are assumed to be able to solve current challenges such as medical issues, *e.g.* drug delivery, diagnostics, as well as energy issues, *e.g.* energy conversion and storage, fuel cells, power consumption reduction, nutrition, *e.g.* food, clean water, electronics, and communication, *e.g.* molecular electronics, and customer goods. Nevertheless, nanotechnology and the accompanying effects on the environment have to be fully understood in order to make a suitable comprehensive risk analysis.

In the last decade, the semiconductor industry has developed as predicted by Moore's law, *i.e.* the number of components in integrated circuits doubles every 18 months.[1] The top-down approach now encounters its limits at an increasing growth rate,[2] *i.e.* the common lithographic techniques face physical,

Motivation

technical, and economical limitations.[3] To fabricate even smaller structures, a new strategy, the bottom-up approach, has been established, which operates at the ultimate length scale of the order of single molecules and atoms, respectively. When operating in this dimension, also quantum phenomena have to be taken into account,[4] because the classical descriptions of processes hit more often the dead end and new theories have to be established.

To investigate nanometer sized structures, scanning probe microscopy (SPM) can be applied and has proved suitable in probing the world of atoms and molecules. In the 1980's, scanning tunneling microscope (STM)[5] and atomic force microscope (AFM)[6] were the first representatives of these analytical tools. The STM utilizes an electric current between a sharp metal tip and the sample. The AFM on the other hand detects a force between a cantilevered sensor and a sample, which makes it applicable to insulating samples. Nowadays, several modifications of these analysis techniques such as scanning capacitance microscopy (SCM), scanning near-field optical microscopy (SNOM), or spin polarized scanning tunneling microscopy (SPSTM) have been established. Most of these methods can be applied under ultra-high vacuum conditions, ambient conditions as well as at the liquid-solid interface. These surface science analysis techniques allow to study a large range of phenomena from structures on the micrometer range down to the sub-nanometer scale: surface topography and structure dimension,[7–9] electronic and vibrational properties,[10] film growth,[11] studies of lubrication and friction,[12,13] determination of adhesion and strength of individual chemical bonds,[14] controlling the charge state of individual atoms,[15] dielectric and magnetic properties,[16,17] contact charging, and molecular manipulation.[18] Nevertheless, it is still an open question whether SPM methods have the capability of being a useful fabrication tool in nanotechnology despite the extreme slowness of this serial approach.

For the creation of novel structures, manipulation[19–21] and imaging of single atoms or molecules[22] is an essential part. The used building blocks, also referred to as tectons,* are often functionalized molecules, inhibiting various chemical functionalities and thus electronic properties.[23] Following the bottom

---
*derived from Greek word τεκτων meaning "from the base of".

up strategy, these building blocks arrange on 2D surfaces to patterns driven by the interplay of attractive and repulsive interaction. In 1987, the Nobel Prize in Chemistry was awarded jointly to Donald J. Cram, Jean-Marie Lehn, and Charles J. Pedersen "for their development and use of molecules with structure-specific interactions of high selectivity". This toolbox of supramolecular chemistry[†] offers a broad perspective on applications in various fields related to nanotechnology *e.g.* molecular information storage devices,[24] self-healing materials,[25,26] functionalized organic surfaces[27,28] such as corrosion resistance,[29] oxidation protection,[30] and organic thin film transistors for large area electronics.[31] Other applications aim at catalysis,[32,33] molecular electronics,[‡][35] electronic transition to organic layers, harvesting sunlight[36], emitting light,[37] nano-machines,[38] artificial muscles[39] and battery materials for ultrafast charging and discharging.[40] For many of these applications, molecules have to be arranged in a repetitive structure and also have to be addressable and manipulable in a controlled manner.[24]

Molecular self-assembled networks that are stabilized by reversible interactions such as hydrogen-bonds and van der Waals forces bear the potential to form complex sophisticated structures with a low defect density. Moreover, the well-known bulk metal coordinated complexes were transferred to two dimensions. The strength of this type of interaction lies in an intermediate position between hydrogen-bonded networks and covalent organic frameworks, albeit these bonds of metal-organic frameworks are reversible under certain conditions.[28] In contrast, the final structure of 2D covalent networks are affected by a high defect density.

This fundamental research work, where some newly developed technical devices are supplemented, should at least increase the understanding of the important parameters for the formation of 2D nanostructures such as self-assembled monolayers (SAMs), metal-organic frameworks (MOFs), as well as covalent organic frameworks (COFs). Hopefully, the gathered knowledge serves as basis for new devices. Apart from the involved forces, *i.e.* molecule-molecule

---

[†]chemistry beyond the molecules, dealing with structures that are composed/assembled of a discrete number of components or molecular subunits.
[‡]Molecular electronics is believed to be the successor of today's electronics based on silicon technology.[34]

and molecule-surface interaction, kinetic and thermodynamic considerations have to be taken into account to understand the complex processes during the synthesis of such networks.[41] The molecules investigated here were mostly composed of π aromatic systems, rendering them suitable for molecular electronic devices due to their conductive organic backbone. Frequently, the terminating lobes were functionalized either with carboxylic groups, halogens, or thiols groups. To form laterally interlinked 2D structures from a single tecton, their symmetry has to be at least threefold. Moreover, the used building blocks tend to adsorb in a planar orientation on the surface at specific adsorption sites.[42] This process of physisorption or chemisorption is dependent on the characteristics of the surfaces, i.e. surface orientation and material. The final structure is also decisively influenced by the reactivity and catalytic properties of the surface. For structures on inert graphite surfaces mostly self-assembly occurs, however, adsorption on transition metals, e.g. Au, Ag, Cu, can facilitate chemical reactions. Furthermore, external parameters such as solvent, concentration and temperature, to name but a few, have a significant influence.

This thesis is organized as follows: Chapter 2 gives a short overview on the fundamental principle of the scanning tunneling microscope, the analysis device which was primarily used in this thesis. At first, the fundamental principle of the device is described followed by a theoretical description of the tunneling process through molecules adsorbed on different surfaces. Finally, the instrumentation and design concept of the STM is briefly presented. Chapter 3 introduces the experimental details and materials used in this work. After a short comparison of the influence of UHV and ambient conditions to the formation of 2D networks, the focus lies on probe manufacturing as this is critical for atomic and submolecular resolution. Following this, the used substrates, solvents, and molecules are introduced. This chapter ends with a description of surface and sample preparation protocols and the devices involved in that. Chapter 4 outlines the concept of self-assembly and self-organization and considers Gibbs free energy, transition state theory, and reveals external parameters governing the formation of 2D networks. Molecule-molecule interaction and interaction of molecules with the substrate, i.e. the processes of physisorption and chemisorption are also discussed. In here, general trends are outlined rather than a concise description is provided. Chapter 5 presents summaries of further unpublished results and published articles. The first summary deals with the incorporation dynamics of host-guest systems using the example of coronene adsorption in a trimesic acid host network. In the second one, the competition of halogen bonded networks versus hydrogen bonded networks is discussed. Third, the intriguing improvement to the imaging quality, *i.e.* submolecular resolution, of networks on graphene terminated substrates is examined. Fourth, the synthesis of metal-organic frameworks on the basis of thiolated building blocks is presented. Fifth, the formation of covalent organic networks employing a radical addition reaction under ambient conditions as well as ultra-high vacuum condition is analyzed. Chapter 6 summarizes the results and gives an outlook. Chapter 7 presents the published manuscripts.

CHAPTER 2

Introduction to scanning tunneling microscopy

This chapter deals with one of the most ubiquitous tools in surface science, the scanning tunneling microscope (STM). This device is commonly used for imaging molecular and even atomic structures on surfaces in real space, in contrast to k-space methods such as diffraction techniques. Section 2.1 introduces the fundamental principle of scanning tunneling microcopy. The following section, 2.2, focuses on the mechanism of tunneling through molecular monolayers on different substrates and elucidates the underlying fundamental physical principles. Section 2.3 presents a technical realization of an STM and provides some details on STM data evaluation.

2.1 Fundamental principle

Besides the atomic force microscope (AFM), the scanning tunneling microscope (STM) has become one of the most important instruments in surface science and nano science.[5,43] In 1981, Gerd Binnig and Heinrich Rohrer succeeded in the technical realization of a device measuring the tunneling current between a sharp tip and a conductive surface while raster-scanning. They were awarded with the Nobel Prize in physics in 1986 together with Ernst Ruska. The first great achievement of their near field microscope was the imaging of a 7×7 reconstruction of Si(111).[44] Later, the STM also turned out to be suitable for positioning[45] and manipulating atoms and molecules.[19–21] In 1991, Eigler *et al.* published an STM topography image of single Xe atoms for the first time.[46,47] Two years later, Crommie and Eigler arose again much public interest by publishing their ultra-high vacuum STM results on a quantum corral, formed out of 48 iron atoms on a Cu(111) surface.[48] Meanwhile, specialized STM setups also enable studies of dynamic surface processes by means of fast image recording as demonstrated by Wintterlin *et al.* for studying atomic equilibrium fluctuations of adsorbed oxygen on Ru(0001)[49] or Tansel *et al.* for sulfur on Cu(100) electrodes.[50]

Providing an exact theoretical description of the imaging process is almost impossible due to experimentally inaccessible information on the electronic states of both the tip and the surface. The theory presented here is deliberately kept short. For further reading, the text books of Lüth[51] and Chen[52] and the references cited therein are recommended.

The operating principle A sharp metal tip is brought in close proximity to the sample surface, *i.e.* a distance of a few Ångströms.[53] In order to achieve a directed net tunneling current I_T of typically 10 pA to 10 nA, a convenient operating bias voltage V_T in the region of ± 2 mV to ± 2 V is applied between tip and the sample.[54] The STM images are recorded by raster-scanning this sharp tip across the surface and detecting the current by means of a highly sensitive transimpedance amplifier. Typically, the control unit is able to operate the STM in two different modes, *i.e.* the constant-height mode and the constant-current mode. To attain the latter, the tunneling current is kept constant by means of a feedback circuit which controls a piezo actuator. The topography images are composed of maps of the displacement of the vertical piezo element at every lateral position. The data is derived from the variation of the piezo voltage $U_z(x,y)$, which adjusts the tunneling current to its set-point value, while the tip is following the surface corrugation. In contrast, the constant-height mode allows much faster imaging due to deactivated feedback control. Hence, the information is contained in a spatial map of the tunneling current signal $I_T(x,y)$. Independent of the chosen mode, the lateral movement of the tip is conducted by utilizing the inverse piezoelectric effect. In order to resolve atoms, the movement in the lateral direction must be controlled with a precision of 0.05 Å to 0.10 Å, which is feasible by piezo ceramics.[54] The obtained images reflect the complicated convolution of the electronic properties of surface and the tip. Detailed information of the technical realization on different setups are presented in section 2.3.

The origin of the tunneling current The fundamental principle of scanning tunneling microscopy is based on the tunneling effect, which allows electrons to penetrate through classically impenetrable potential barriers. Figure 2.1 shows the schematic of the electron tunneling process from a metal tip through the vacuum gap into a metal surface, where the electrons in the nth sample state, *i.e.* the quantum number of the energy levels, are represented by their wave functions Ψ_n.

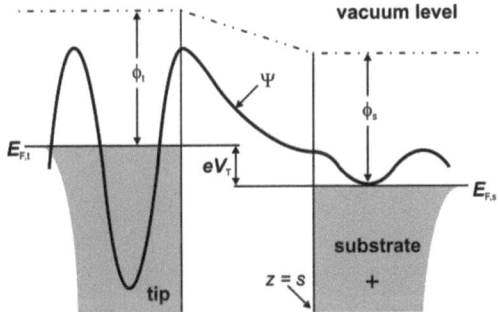

Figure 2.1: Sketch of a tunneling junction including a one dimensional potential barrier between two metal electrodes, where the electron is presented with its wave function Ψ_n. A negative tunneling voltage V_T is applied, *i.e.* the electrons tunnel from occupied tip states into empty sample states. The Fermi levels of sample $E_{F,s}$ and tip $E_{F,t}$ and the related work functions ϕ_s and ϕ_t are indicated with respect to the vacuum level.

The probability w for an electron to be present at the sample surface, *i.e.* $z = s$, is

$$w \propto |\Psi_n(0)|^2 \, e^{-2\kappa s} \qquad (2.1)$$

where κ is the decay constant which is defined as $\kappa = \sqrt{2m\phi}/\hbar$ with m being the electron mass, ϕ the work function,* and $\hbar$ reduced Planck's constant ($\hbar = \frac{h}{2\pi}$). For simplicity, it is assumed that the work function of tip and surface are equal. In order to achieve a net tunneling current, a small bias voltage ($eV \leq \phi$) is applied. For electrons in the energy interval between $E_F - eV$ and E_F, there is a non-vanishing probability for tunneling.[52] Yet, this tunneling current I_T is proportional to the sum over all contributing states. Therefore, it can be stated that

$$I_T \propto \sum_{E_n=E_F-eV}^{E_F} |\Psi_n(0)|^2 e^{-2\kappa s} \qquad (2.2)$$

For sufficiently small bias voltages Equation 2.2 can be conveniently written in terms of the the local density of states (LDOS, ρ_s) at the Fermi level

$$\rho_s(z, E_F) \equiv \frac{1}{eV} \sum_{E_n=E_F-eV}^{E_F} |\Psi_n(z)|^2 \qquad (2.3)$$

The LDOS is a physical quantity that gives the space-resolved number of electrons per unit volume per unit energy at a given energy. Substituting Equation 2.3 into Equation 2.2 yields

$$I_T \propto V \rho_s(0, E_F) e^{-2\kappa s} \qquad (2.4)$$

Hence, the tunneling current depends exponentially on the sample-probe distance, which makes the instrument utmost sensitive to corrugations of the surface electron density. For typical experimental imaging conditions, an increase of the sample-tip distance of about 1 Å already causes a reduction in I_T to approximately one-tenth.

Bardeen formalism, Tersoff-Hamann-Model In 1961, two decades before the invention of the STM, Bardeen used time dependent first order perturbation theory to calculate the tunneling current between two planar metal plates.[55]

---

*the minimum energy, which is required to remove an electron from the bulk to the vacuum level, depending on the material and the crystallographic orientation of the surface.

Using this approach, he avoided solving the Schrödinger equation of the combined system. Instead, Bardeen solved the Schrödinger equations for two unperturbed subsystems. The electron transfer between the two electrodes, *i.e.* the probability of transition Γ, is determined by Fermi's golden rule and expressed as

$$\Gamma = \frac{2\pi}{\hbar} |\boldsymbol{M}_{\mu\nu}|^2 \delta(E_\mu - E_\nu) \qquad (2.5)$$

where $\boldsymbol{M}_{\mu\nu}$ is the tunneling matrix between the states ψ_μ of the probe and ψ_ν of the surface. E_μ and E_ν are the energies of state ψ_μ and ψ_ν, respectively, in the absence of tunneling.[56]

The general problem in handling Equation 2.5 is to calculate $\boldsymbol{M}_{\mu\nu}$, which requires the knowledge of the wave functions of the electrodes. Bardeen showed that the tunneling matrix element $\boldsymbol{M}_{\mu\nu}$ is determined by a surface integral on a separation surface A between the states of the probe ψ_μ and the surface ψ_ν:[52]

$$\boldsymbol{M}_{\mu\nu} = -\frac{\hbar^2}{2m} \int_A \left(\Psi_\mu^* \boldsymbol{\nabla} \Psi_\nu - \Psi_\nu \boldsymbol{\nabla} \Psi_\mu^* \right) d\vec{A} \qquad (2.6)$$

Ψ_μ is the wave function and the area integral is over any arbitrary surface in the barrier region.[54] J. Tersoff and D. R. Hamann modeled the tip as a radially symmetric wave function in order to provide a framework for interpreting scanning tunneling microscope data. Up to now, the Tersoff-Hamann Model is the standard model for three dimensional tunneling, despite several oversimplifying assumptions.[56,57] In Bardeen's formalism, the tunneling current, *i.e.* $I_T = e\Gamma$, is given by

$$I_T = \frac{2e\pi}{\hbar} \sum_{\mu,\nu} f(E_\mu)[1 - f(E_\nu + eV)] |\boldsymbol{M}_{\mu\nu}|^2 \delta(E_\mu - E_\nu) \qquad (2.7)$$

Taking only low temperatures into account, the Fermi function can be approximated by a step function. Due to restriction of the bias voltage to small values and elastic tunneling, *i.e.* $E_\nu = E_\mu$, only states in immediate vicinity to the Fermi level contribute to the tunneling current. Equation 2.7 is therefore reduced to

$$I = \frac{2\pi}{\hbar} e^2 V \sum_{\mu,\nu} |\boldsymbol{M}_{\mu\nu}|^2 \delta(E_\nu - E_\mathrm{F}) \delta(E_\mu - E_\mathrm{F}) \tag{2.8}$$

By the assumption of a localized s-wave function as the probe and a Bloch wave decaying exponentially perpendicular to the surface, the tunneling current is proportional to the amplitude of Ψ_ν at the center of the spherical potential $\vec{r_0}$ of the probe.[56] For this reason Equation 2.8 can be simplified to

$$I \propto \underbrace{\sum_\nu |\Psi_\nu(\vec{r_0})|^2 \delta(E_\nu - E_\mathrm{F})}_{\rho(\vec{r_0}, E_\mathrm{F})} \tag{2.9}$$

Consequently, the tunneling current in STM imaging displays the contour of the surface local densities of states (LDOS, $\rho(\vec{r_0}, E_\mathrm{F})$) at the Fermi energy at the position, where the center of the spherical tip is located.[56] However, this s-wave-tip model is not suitable to explain atomic resolution of metal surfaces as well as the imaging of adsorbed molecules, but it provides a qualitative picture of the surface.[52] For a more quantitative understanding of the imaging process a more general theory is needed.[58] Such approaches will be described below.

Further theoretical approaches The following models for calculating the tunneling current are only introduced qualitatively. A detailed overview is presented in the work of G. A. D. Briggs and A. J. Fischer[59] and references herein.

N. D. Lang computed the tunneling current of an additional adsorbed atom between two planar metal electrodes in his pioneering work.[60] He modeled both electrodes with a jellium model[†] and took the effects of adatom valence resonances entirely into account. The resulting current density was derived in analogy to Bardeens approach and enabled the model to determine the contrast of single atoms adsorbed on a metal surface in STM images.[61] Eigler *et al.* demonstrated the applicability of the atom-on-jellium model by comparing the

---

[†]uniform electron gas model, quantum mechanical model of interacting electrons, which are assumed to be uniformly distributed in space.

measured apparent height of a Xe atom with its calculated value, which were in good agreement.[47]

For all the improvements considered by Lang, the atomic structure of the tip remained still disregarded. C. J. Chen extended the Tersoff-Hamann s-wave tip approach by considering the detailed structure of the tip's wave functions, *i.e.* a linear combination of the wave function and its (first and higher) partial derivatives. Accordingly, the electronic structure of the tip was assumed as localized surface states, for example metallic surface d_{z^2} states (*e.g.* W, Pt, and Ir) or p_z dangling-bond states (*e.g.* silicon).[62] Therefore, the nucleus of the apex atom of the tip tracks the contour, which is determined by the derivatives of surface wave functions of the sample, causing a much stronger atomic corrugation than Fermi-level LDOS.[63]

A recent approach to interpret STM images with subatomic resolution was established by P. Sautet in terms of the elastic scattering quantum chemistry (ESQC). This approach treats the tunneling process as a scattering phenomenon with a defect at the position of the tip apex-adsorbate-substrate tunneling junction.[64,65] ESQC exceeds the limitation of theories based on electron-density maps or the Tersoff-Hamann local density of states formula which are all based on the Bardeen formalism that is restricted to short tip-substrate distances. In ESQC, the entire Hamiltonian for the tunneling junction (tip apex, barrier, and surface) is formed. The conductance between tip apex and substrate surface, that are both connected to an electron reservoir, is calculated by an generalized Landau formula.[64] The required elements of the scattering matrix are extracted from a quantum chemistry calculation.[66]

Due to raising computational power, even more complex numerical calculation can be conducted including atomic shape of tip and surface and adsorbed substrates. Nevertheless, all models, whether the class of perturbation theory or the class of scattering theory formalism, depend crucially on the assumptions made, *e.g.* that tunneling does not change the properties of the system.

2.2 Tunneling through molecules – Contrast mechanism and interpretation

Once the STM data are acquired, the major challenge is to match the obtained pattern with a proposed model based on the chemical structure of the adsorbed molecules. The resulting topographs convolute the geometric and electronic properties of both sample (molecules and surface) and tip. As a consequence, the attained data are affected by the respective materials of tip and surface, the associated chemical and electrostatic interaction, the material inherent electronic properties, the applied bias voltage, and the distance between tip and surface.[18] Datta *et al.* described the tunneling through molecules by a scattering theory of electron transport, taking into account the chemical potential of the electrodes involved.[67]

The intuitive interpretation of STM images as a direct measure of topological height is not generally correct, in particular for adsorbates on metal surfaces. For instance, oxygen atoms adsorbed on metal surfaces appear as depressions even if they sit on top of the bare metal surface. Thus, knowledge of the relative alignment of energy levels of the involved molecules or atoms is crucial.[68] Under tunneling conditions, an electron has a non-vanishing probability to pass the potential barrier between the two "electrodes" forbidden in the classic picture. This basic model is expandable to adsorbed atoms and molecules on surfaces. To describe the tunneling process, the theory introduced in section 2.1 has to be extended by an additional component which considers the newly formed tunneling junction (see Figure 2.2).

Basically, the electronic coupling Γ_C of the adsorbed molecules to the surface is expected to be stronger in comparison to the coupling to the STM tip.[67] The prerequisite for imaging partially electrically insulating molecules on top of a conductive surface is sufficiently thin layers. This limitation is caused by the exponential distance dependence of the tunneling current on the tip-sample separation. The packing density of the molecules on the surface, *i.e.* the surface coverage, can also influence the observed contrast by molecule-molecule interactions.

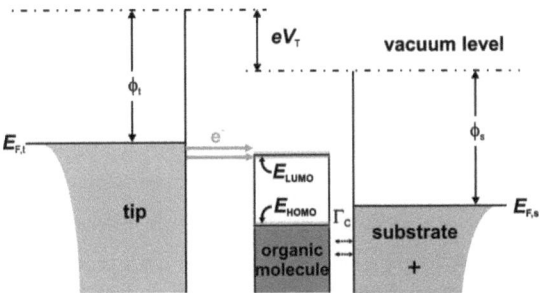

Figure 2.2: Energy diagram for a typical tunneling process through an organic molecule for a negative tunneling voltage V_T. The position of the frontier orbitals (HOMO and LUMO) and the Fermi level of the sample $E_{F,s}$ and the tip $E_{F,t}$ are indicated. Γ_C describes the electronic coupling between adsorbate and substrate, which is the cause of broadening (shaded blue) and shifting of the molecular energy levels.[69]

It is generally accepted, that the STM contrast is governed by the applied bias tunneling voltage V_T and in particular its polarity.[‡] By switching the sign of the tunneling voltage, electrons can tunnel either into the lowest unoccupied molecular orbital (LUMO) for negative tunneling voltages (see Figure 2.2) or out of the highest occupied molecular orbital (HOMO) for positive tunneling voltages. Hence, different features of the same molecule or sample can emerge in the STM topographs, depending on the predominantly involved molecular orbitals. Lackinger et al., for instance, observed a dark region representing the center of naphthalocyanine (Nc) molecules in a bias dependent study of Nc molecules on graphite, when measuring with a negative bias voltage.[70] This indicates tunneling out of occupied states. For positive bias voltage, on the other hand, all molecules appear as bright spots indicating tunneling into unoccupied states.

In addition to the polarity of the bias voltage, which selects whether tunneling into LUMO or out of HOMO, the magnitude of the bias voltage can be identified as the reason for varying image contrast of the molecules under investigation.[71] Typically, the applied bias voltages in STM experiments, which are assumed to drop entirely at the tunneling barrier, are well below the spectroscopic and

---

[‡]To avoid confusion, a positive sign of V_T means tunneling into the tip. Therefore, a positive tunneling voltage is equivalent to a negative sample bias voltage: $V_T = -V_S$.

electrochemically band gap of 5 eV to 10 eV that characterize the molecules.[53] The physical reason for the occasional occurrence of STM contrast variation can be based on the injection of the electrons into different molecular orbitals (HOMO, HOMO^{-1}, HOMO^{-2}, LUMO, LUMO^{+1}, LUMO^{+2}) depending on the applied bias voltage. Figure 2.3 (a) - (c) presents a bias voltage height dependence study of a single coronene molecule adsorbed in a trimesic host matrix on graphite. A line profile through the occupied cavity is shown below the respective STM topographs.

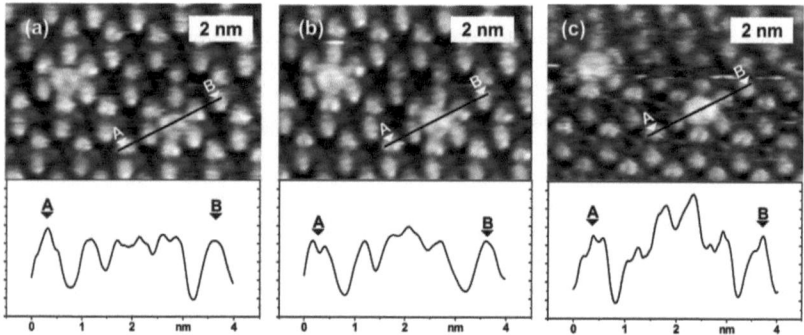

Figure 2.3: Bias voltage dependent study for a single coronene molecule adsorbed in a cavity of a trimesic host network structure. The STM topographs (a) - (c) show the same area scanned with different voltages, i.e. (a) $V_T = 0.76$ V, (b) $V_T = 0.96$ V, and (c) $V_T = 1.26$ V whereas the lower part of the images present a cross section along the indicated lines. All other parameters were kept constant. STM tunneling conditions: $I_T = 53.5$ pA.

The higher the applied voltage the higher the coronene molecules appear in contrast to the trimesic molecules. In order to approach a more precise height-value for coronene, the known lattice constant of a step-edge on the substrate can be utilized as a rough calibration standard.

Nevertheless, the contrast of STM topographs of organic molecules adsorbed on surfaces is regularly independent over a certain range of bias voltage. Lippel et al. explain this behavior by close resemblance of HOMO and LUMO bands of the investigated molecules.[72]

In the case of liquid-solid experiments, the presence of any solvent affects the electric field in the vicinity of the STM tip and can lead to a shift of energy levels accompanied by a change in STM contrast.[69] Basically, molecules adsorbed on surfaces are able to modify the local density of states (LDOS) at the Fermi level and therefore affect the STM contrast. Molecular orbitals, whose energies are close to the Fermi level, have a stronger contribution to the tunneling current and appear brighter than those far away from the Fermi level.[73] In general, brighter shading in STM images is addressed to higher conductivity. Moreover, modification of the probe, *e.g.* picking up a molecule by the tip apex, during the scanning process can facilitate different tunneling conditions, which may result in an inverted STM contrast. In order to change the tunneling conditions and modify the tip, a short voltage pulse can be applied. In addition, the chemical conformation of the adsorbed species and the adsorption sites also plays an important role on for the STM contrast.[74]

Claypool *et al.* presented an explanation of STM contrast at submolecular resolution observed for functionalized alkanes and alkyl alcohols, taking the ionization potential of the functional groups into account.[75] Identification of individual hydrogen atoms or the determination of the length of methylene units (bright spots in STM contrast) at least was possible in most cases. De Feyter *et al.* described contrast features of physisorbed organic monolayers (derivates of isophthalic and terephthalic acid) on highly oriented pyrolytic graphite.[76] As a rule of thumb, the authors observed that aromatic moieties, for instance phenyl rings, are imaged as bright spots, whereas the darker regions correspond to the interdigitated alkyl chains.

Another phenomenon occurring in STM images of ordered monolayers adsorbed on crystalline surfaces is the observation of a large scale contrast modulation.[77] The so called Moiré patterns§ are reflected in a slight periodic variation in the apparent height of STM topographs. Prerequisite for the appearance of Moiré pattern in organic (mono)layers on crystalline surfaces is an incommensurate relation between the lattice of the adsorbate film and the substrate giving rise to this superstructure. In contrast to Moiré patterns occurring in transmission electron microscopy (TEM), Moiré pattern in STM images cannot be simply

---

§Moiré pattern occur in different places in nature as a result of superposition of two slightly different lattices generating a third.

understood in terms of superposition of two slightly different lattice constants. In STM, these patterns originate primarily from three-dimensional tunneling and the fact that nanoscale waves, generated by the interface scattering in lattice-mismatched systems, propagate through many layers without decay.[78] Due to the high sensitivity of Moiré patterns, Diaye et al. used such superstructures to determine epitaxial relations between graphene and Ir(111).[79]

Figure 2.4 (a) presents an STM topograph of a trimesic acid (TMA) dissolved in nonanoic acid on highly oriented pyrolytic graphite after the self-assembly process was finished. A detailed view on the STM topography reveals two different lattices occurring in this honeycomb structure, i.e. the hexagonal trimesic acid network as well as the TMA superstructure lattice. In order to determine the exact lattice parameter, fast Fourier transformation (FFT) tends to be appropriate as done in Figure 2.4 (b). The spots in the FFT spectra were addressed either to the TMA network lattice or the Moiré Pattern and they were measured to be (1.6 ± 0.1) nm and (5.8 ± 0.1) nm, respectively. On basis of FFT spectra analysis a rotation by $(10 \pm 1)°$ of the TMA superstructure lattice with respect to the TMA lattice was also assessed, which is in good agreement with a previous study.[42] In general, for determining the lattice parameters as well as the relative (epitaxial) orientation with respect to each other, the evaluation of a FFT spectra is the method of choice rather than the evaluation of the respective STM topographs.

Another important factor that determines the contrast of STM topographs is the substrate on which the molecules are adsorbed. Normally, STM studies have been conducted either on metal substrates or on graphite. Recently, graphene terminated substrates (e.g. conducting Cu foil or insulating SiO_2, for details see section 3.3) are used to investigate various atoms and organic molecules. The behavior of the adsorption and desorption process of molecules on graphene terminated substrates seems to be similar to graphite.[80] Due to the unique properties of graphene, i.e. that the electrons within a layer can roam freely for hundreds of nanometers before they hit a defect,[81] submolecular resolution for conjugated π electron systems, that are decoupled from graphene or the

---
¶Publication in preparation: Georg Eder, Izabela Cebula, and Peter Beton, University of Nottingham.

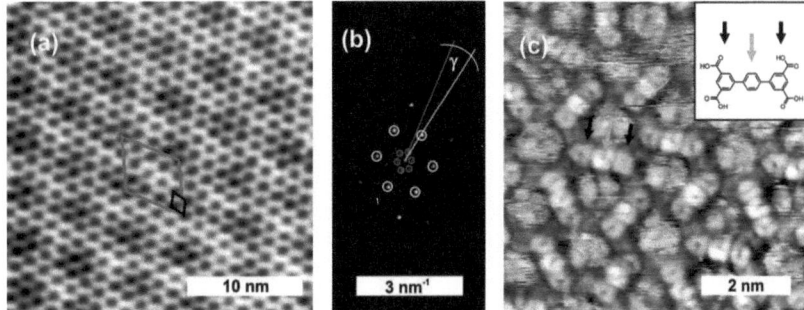

Figure 2.4: (a) STM topography of the TMA honeycomb self-assembled monolayer on highly oriented pyrolytic graphite. (b) 2D fast Fourier transformation spectra of the raw data. The inner spots (encircled red) correspond to the TMA superstructure network, *i.e.* the Moiré pattern, whereas the outer white encircled spots are related to the TMA network. The angle γ indicates a rotation of the TMA network relative to the TMA superlattice of approximately $(10 \pm 1)°$. (c) STM topography of p-terphenyl-3,5,3',5'-tetracarboxylic acid (TPTC) molecules adsorbed on graphene terminated Cu foil.¶A chemical model of the molecule is presented in the inlay. The blue arrow marks the brighter inner phenyl ring, whereas the black arrows highlight the outer phenyl rings. STM tunneling conditions: (a) $V_T = 0.50\,\text{V}$, $I_T = 48.1\,\text{pA}$; (b) $V_T = 0.60\,\text{V}$, $I_T = 50.0\,\text{pA}$.

underlying substrate were reported.[82] In contrast to the adsorption on graphite, there is less contribution from the bulk material and the aromatic molecules couple with the p_z orbitals to the graphene. The incredible sensitivity of this substrate is demonstrated in Figure 2.4 (c), where individual phenyl rings are resolved on the nonanoic acid - graphene (on Cu foil) interface. The single aromatic rings of one p-terphenyl-3,5,3',5'-tetracarboxylic acid (TPTC) molecule are exemplarily marked with color coded arrows. Mostly, the inner phenyl ring appears brighter. Identical high resolution topographs of the same system were obtained on a graphene sheet, terminating an insulation SiO_2 substrate.

2.3 Instrumentation[||]

The scanning tunneling microscope provides the technical platform for imaging atoms by exploiting the tunneling effect for electrons. Therefore, the designs primarily have to provide stability in order to avoid drift in vertical and lateral directions during raster-scanning of the tip. Within the last 30 years, a variety of different STM designs (*e.g.* louse[5], Besocke,[83] or possible variation such as Frohn[84] or Wilms[85]) have been realized, operating under different environmental conditions such as ultra-high vacuum,[86] high pressure,[87] liquids,[88] or even under high magnetic fields.[89] Some STMs have also been designed for operating at extreme temperatures, *i.e.* from extremely low temperature of 30 mK[90] to extremely high temperatures of up to 1100 K.

The fundamental operating principle of the scanning tunneling microscope is simple. However, the components have to meet high requirements in construction and design. In the following, the primarily used setup in this work is described. It can be operated in different modes, *i.e.* as scanning tunneling microscope (STM) and non-contact atomic force microscope (NC-AFM). Figure 2.5 presents an overview of the major assembly parts. The device was designed and realized by Stephan Kloft.

The metal tip is clamped by a set-screw in the magnetically attached STM probe holder ④. A piezo tube scanner ② enables the raster-scanning across the substrate, which is electrically connected to the sample ⑤. A permanent magnet attaches the sample to the base plate ⑥, which renders the setup very versatile in terms of an adaptable sample holder. A linear positioner ① is utilized for the coarse approach. The whole STM head ③, including the electrical wiring, rests on three magnetic feet on the base plate. The coarse approach mechanism is based on an interplay between the voltage driven expansion and contraction of the scanner tube and the movement of the whole scanning unit, driven by a modified linear positioner (a modified ANPz50**) utilizing the stick-slip effect.

[||]Publication in preparation:"A versatile drift-stable Scanning Probe Microscope with inertial coarse approach for molecular self-assembly studies", Stephan Kloft, Georg Eder, Khaled Karrai, Wolfgang M. Heckl, and Markus Lackinger.
**attocube systems AG, Königinstrasse 11a RGB, 80539 München, Germany.

Introduction to Scanning Tunneling Microscopy

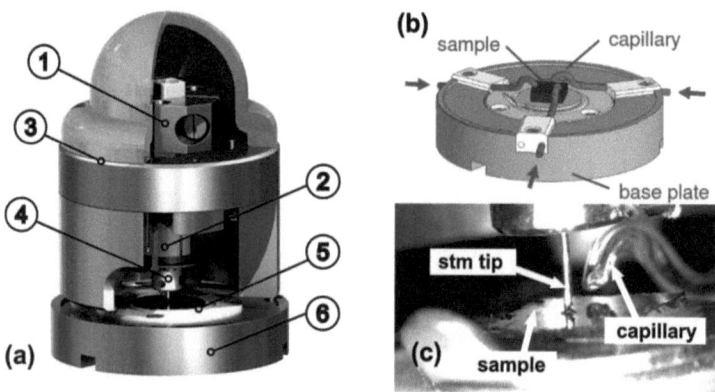

Figure 2.5: Sketch of the setup of the scan-head on the sample platform with its major assembly parts: ① attocube linear positioner, ② piezo tube, ③ housing/shielding, ④ probe holder, ⑤ sample platform with sample-pad and sample-holder, ⑥ base plate. CAD drawing (b) and photograph (c) of the injections add-on which enables the application of additional solvent during raster-scanning.

First, the axis of the positioner performs a slow acceleration, leading to a mass transport by static friction. This movement is followed by a fast acceleration in the opposite direction, which causes the positioner to slip over the clamped axis. A periodic repetition results in a stepwise motion that is controllable with nanometer precision over a millimeter range. A combination of this step-by-step motion with the expansion and contraction of the tube scanner piezo is repeated until tunneling contact is established.

The tube scanner is intended for movement in the $x-y$ plane planar-parallel to the sample surface as well as for z-positioning of the tip in the vertical direction. The latter is achieved by a closed loop feedback. The tube scanner used in this setup has four outer segments and one inner segment for electrical supplies. This type was realized by Binnig and Smith for the first time.[91] Applying voltages on the opposite outer segments results in the contraction of one and the expansion of the opposite side, and thus, an x/y positioning by bending of the tube. The vertical motion is obtained by applying a voltage to the inner electrode with respect to all four electrodes on the outside.

For studing the dynamics of guest incorporation in a pre-existing host matrix at the liquid-solid interface a novel injection add-on (see Figure 2.5 (b) and (c)) was developed.[92] This system facilitates data acquisition while applying additional solution through a bent glass capillary, which is directed at the sample. The injection of the solution takes place under visual camera control by means of a thoroughly mechanically decoupled syringe outside the shielding of the STM. The injection neither impairs high resolution nor caused substantial drifts.

Besides the mechanical components of the STM, the characteristics of the used electronics regarding amplification, bandwidth, reaction time, stability and signal to noise ratio (SNR) play a major role. To date, special control systems allow frame rates up to 200 Hz.[93] Figure 2.6 depicts a schematic plan of a standard STM setup consisting of a scanning unit and a control electronics. The separation of signal and high voltage supply wiring is introduced to avoid possible crosstalk. In addition, an outer housing shields the instrument from electromagnetic fields. Besides the feedback controller,[††] the high-voltage amplifier, a very sensitive current-to-voltage converter (IVC, typical bandwidth 1 kHz) and a software interface complete the whole setup.

To achieve high resolution STM images or to perform scanning tunneling spectroscopy (STS) measurements, the noise of both mechanical and electronic origin have to be minimized or – ideally – reduced to their physical minimum. An extensive contribution to mechanical noise reduction can be achieved by mounting the vibration-isolated STM setup on a damped optical table, which suppresses much of environmental vibrations. The resonance frequency of the custom built damping stage is in the range of 2 Hz to 3 Hz and its transfer function shows excellent damping characteristics for higher frequencies. However, the internal resonance frequency of the scanning head is in the range of 1.7 kHz. This is in very good agreement with the requirement of vastly differing eigenfrequencies of the damping stage and the supporting table.[94]

To reduce the aftereffects of thermal drift during the measurements the choice of construction materials as well as the design details are crucial. In order

---

[††]in most cases a PI controller, $X(t) = g \cdot \left[e(t) + \frac{1}{T_i} \cdot \int_0^\infty e(\tau) d\tau \right]$, $X(t)$ manipulated variable (voltage on the z-direction of the scanner); e control deviation, error; g proportional gain; $\frac{1}{T_i}$ integral gain.

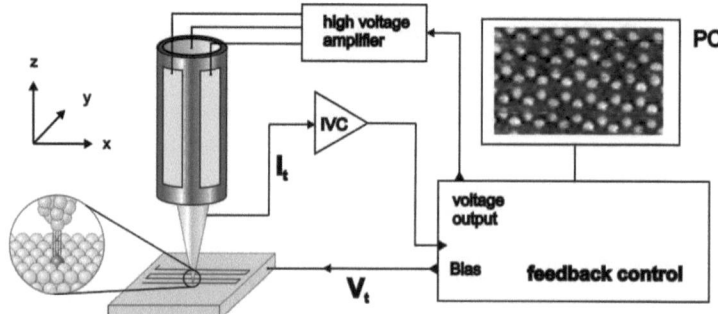

Figure 2.6: Overview sketch of a standard STM setup with its major assemblies: An IVC (current-to-voltage converter) to convert the tunneling current in easier to process voltage, a high voltage amplifier to operate the piezo elements, a feedback control to adjust the tunneling distance in dependence of the selected current set point, and a PC interface.

to reduce these negative impacts, the present symmetrically designed setup was made of Invar steel, which has an extremely low linear thermal expansion coefficient of $1 \times 10^{-6}\,\text{K}^{-1}$ at room temperature.

Data Analysis and evaluation Typically, an acquired STM data set consists of a 512 × 512 pixel matrix initiated by a header, which includes chosen experimental values such as current set-point, bias voltage, and feedback parameters. Each data value is assigned to a physical signal such as tunneling current (current image) or the required voltage for readjusting the tip surface distance (topography image) when running the STM in constant current mode. An image processing software[‡‡] converts this value corresponding to a lookup table (LUT) to false color images. Afterwards, various types of operations can be conducted in the post-processing to enhance the visual representation of the raw data. Besides the basic operations (rotating, flipping, cropping, etc.), data correction functions (filtering, line corrections, *etc.*), transformation functions (2D fast Fourier transformation, 2D FFT) also detections, analysis, and statistical programs can be applied to the data.

[‡‡]commercial image processing software: Scanning Probe Image Processor (SPIP) from Image Metrology, Freeware: Gwyddion, ImageJ.

The evaluation of split images is a common technique to determine unknown lattice parameters of self-assembled patterns and possibly correct images from distortion. This special type of images is recorded by modifying the scanning parameters during the image acquisition and presents the adsorbate structure as well as the substrate within a single frame (see Figure 2.7). After fast Fourier transformation is applied to the raw data, the lattice parameter of the known substrate can be corrected to its real values. This also leads to correct values for the parameter of the unknown structure.

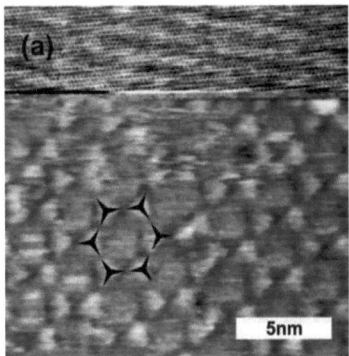 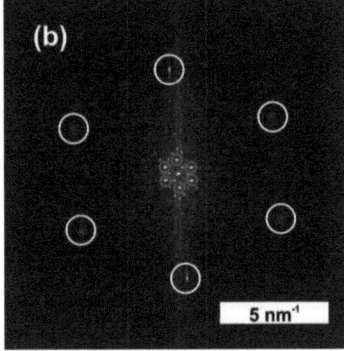

Figure 2.7: Split image of a distorted BTB monolayer on HOPG. (a) The lower part of the STM topography presents the lattice of the self-assembled sixfold BTB network and the upper part the known graphite lattice. (b) 2D fast Fourier transform (FFT) of the split image. The white encircled spots correspond to the substrate lattice. However, the red encircled spots are related to the hexagonal lattice of the organic adsorbate. To eliminate the influence of the drift, the graphite lattice can be adjusted to its known values. The relationship between adsorbate and substrate lattice, and therefore the superstructure matrix, can also be directly derived.

CHAPTER 3

Experimental Details and Materials

This chapter summarizes the experimental details and materials used in this work. In section 3.1, a comparison of ultra-high vacuum conditions versus ambient conditions from the STMs' perspective is presented. Section 3.2 discusses the critical process of probe preparation, which is of pivotal importance to achieve atomic and submolecular resolution. The emphasis of section 3.3 is on the used substrates, solvents, and molecules. Subsequently, the preparation of clean surfaces and the specific sample preparation as well as the used devices are introduced in section 3.4.

3.1 Ultra-high vacuum versus ambient environment

A significant number of the experiments presented in this thesis were performed under ultra-high vacuum conditions at a base pressure of approximately 5×10^{-10} mbar. The used UHV system (see Figure 3.1 (b)) consists of a main chamber and a load-lock offering the possibility to quickly transfer samples and STM tips without venting the main chamber.

The whole framework is mounted on four pneumatic suspension posts facilitating active damping. For surface analysis, the chamber is equipped with an Omicron STM (①), a low-energy electron diffraction system (LEED, ④), and a quadrupol mass spectrometer (QMS, ②). Furthermore, there are several

Experimental Details and Materials

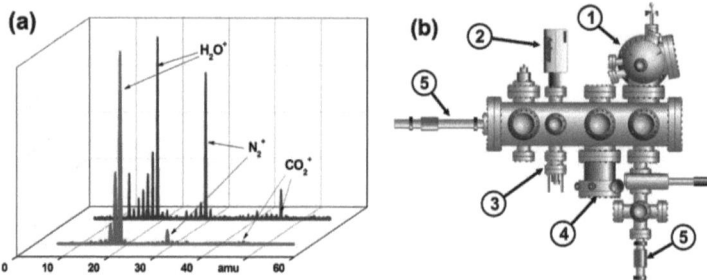

Figure 3.1: (a) The graph shows residual gas spectra from 5 u to 55 u before and after bake-out of a UHV chamber, depicted in arbitrary units for visualization reasons. The difference between the ratio of the H_2O^+ peak and N_2^+ peak before and after the bake-out is obvious. (b) The sketch represents the used UHV chamber equipped with an Omicron VT1000 STM (①), a low-energy electron diffraction system (LEED, ④), and a quadrupol mass spectrometer (QMS, ②), a Knudsen cell (③) and the arms for manipulation of the samples (⑤), adapted from [95].

devices for surface preparation integrated such as an ion-sputtering gun, various evaporation cells (③), and heaters for both sample and tip preparation. The UHV conditions are achieved by oil-free pumps, namely by a combination of a low-vibration magnetically levitated turbo-molecular pump coupled to a fore-pump. An ion pump and a titanium sublimation pump for removal of residual gases are used for the last two orders of magnitude in pressure.

UHV experiments demand ahead-of-schedule work. After an UHV system had been vented, a bake-out (typical temperatures are in the range of 150 °C) of the whole system and its components is essential. The applied heat increases the desorption rate of adsorbed molecules, primarily of water molecules, and therefore facilitates low chamber pressures in acceptable time. Simultaneous degassing of parts exposed to air such as filaments supports re-gaining UHV conditions. A residual gas spectrum before and after the bake-out (see Figure 3.1 (a)) is used as quality criterion. Before bake-out, peaks corresponding to ionized water molecules at 18 u and for its fragments at 2 u, 16 u, and 17 u are dominating the spectra. After bake-out, hydrogen (1 u and 2 u) is the prevailing residual gas besides minor traces of CO, CO_2, and hydrocarbons.[96] A peak at 32 u associated to molecular oxygen would indicate a leak.

UHV conditions provide well-defined and clean conditions for reproducible experiments and are essential for minimizing the complexity and diversity of molecule-surface interactions. In order to achieve less than 10^{13} residual gas molecules per cm^2 on the surface (coverage of roughly 1/100) within a typical time span of an experiment of 10^4 s, a base pressure of 5×10^{-12} mbar is vital.[96] However, for the simple reason that most of the residual gas molecules are harmless to the surface, a base pressure of 5×10^{-10} mbar is sufficient. Moreover, UHV condition enables to control the total number of molecules on the surface by adjusting the molecular flux and taking into account the sticking coefficient, *i.e.* the ratio of the adsorbed molecules to impinging atoms. Based on the controlled way of preparation, the reactions can mainly be attributed to the fundamental reaction principles.* Beyond that, the low number of free residual gas molecules renders thermal treatments possible, neither causing oxidation nor other unwanted processes. Moreover, many surface analysis techniques are based on the excitation with or the detection of electrons which at least demand high vacuum conditions.

Molecules on surfaces under UHV conditions may undergo basic processes such as diffusion, desorption, or even reactions like dissociative chemisorption.[97,98] In addition, using vacuum conditions enables the independent manipulation of parameters, such as the surface temperature or impinging rate of molecules. Due to the low gas density, also molecule-molecule collisions are mostly avoided and the molecules reach the substrate within their initial energy state.

In contrast, under ambient conditions, *i.e.* at the liquid-solid interface, a complex interplay of adsorption and desorption of the investigated molecules and interactions with solvent molecules comes into play. The dynamic exchange of molecules promotes self-repair but also complicates the formation of sub monolayers or even clusters. For self-assembled monolayers at the liquid-solid interface, the finally observed structures are commonly in thermodynamic equilibrium with the supernatant solution. Furthermore, the viscosity of the solvent can influence the kinetics of adsorption. What is more, the concentration

---

*Gehard Ertl also relied on UHV conditions to establish the surface science approach which can play a major role in elucidation of the molecular principles (involved basic physical and chemical processes) that govern reaction on catalyst surfaces.[43]

Experimental Details and Materials

at the interface tends to be a parameter that is challenging to control. Although the evaporation of the solvent can be suppressed, a concentration gradient can hardly be prevented. Moreover, the surface is exposed to a mixture of air and water molecules, depending on humidity. This can also lead to unpredictable conditions at the interface. Moreover, dust particles can also serve as nucleation seeds starting the formation of unwanted networks or phases.

Nevertheless, there were similar or even identical morphologies and structures for self-assembled monolayers (SAM) reported, both for ambient and for ultra-high vacuum conditions.[99,100] For instance, Figure 3.2 (a) shows STM topographs of the self-assembled honeycomb network of 1,3,5-tris(4-carboxyphenyl)benzene (BTB) on graphite at room temperature. The upper part of the STM image presents the self-assembled host structure acquired under ultra-high vacuum after the building blocks were evaporated by means of organic molecular beam epitaxy (OMBE). The same system has been observed under ambient conditions using nonanoic acid as solvent. The resulting structure is displayed with the same length scale in the lower part of the image. Both conditions lead to exactly the same hexagonal host network with an identical lattice parameter of 3.2 nm.[101] They all have in common, that the network is stabilized via double hydrogen-bonds between the building blocks. The registry of the molecules on the surface is not influenced by the environmental conditions.

In contrast, there are several systems, especially ones where surface-mediated reactions are involved, that end up in different structures. Yet, under ambient conditions some types of chemical reactions are blocked or hampered, independent from the used solvent. For instance, 1,3,4-tris(4-bromophenyl)benzene (TBPB) on Au(111) forms a covalent network in UHV after warming the sample (for details see section 5.5). Applying the same protocol to the liquid-solid interface while dissolving the TBPB molecules in ethanol results in just dimers.[102] A practicable approach to overcome this restriction is to transfer the sample after the synthesis under UHV conditions was completed to ambient conditions. Figure 3.2 (b) displays an STM topograph of a TBPB network, which was prepared under UHV conditions and stored for 48 h under ambient conditions before the image was acquired. Obviously, the network is not composed of

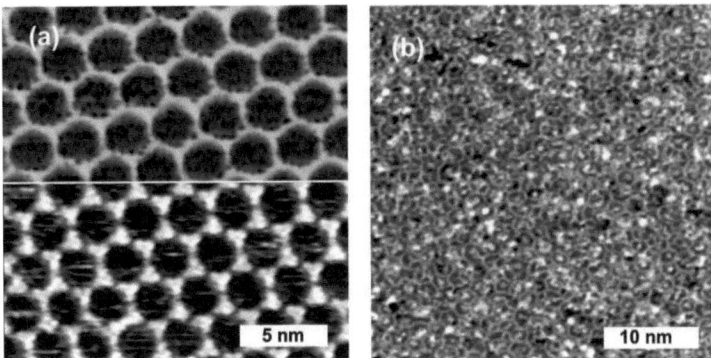

Figure 3.2: (a) A composition of two STM topographs of 1,3,5-tris(4-carboxyphenyl)benzene at different environmental conditions. The upper image was acquired under ultra-high vacuum condition, whereas the lower was imaged at the liquid-solid interface. (b) STM topograph of a 2D polymer (1,3,4-tris(4-bromophenyl)benzene (TBPB)) synthesized under ultra-high vacuum conditions and transferred to ambient conditions. STM tunneling conditions: (a) $V_T = 0.75\,\text{V}$, $I_T = 62.6\,\text{pA}$ for the upper part and $V_T = 0.70\,\text{V}$, $I_T = 55\,\text{pA}$ for the lower part; (b) $V_T = 0.77\,\text{V}$, $I_T = 61\,\text{pA}$.

perfect hexagons and some dust particles are adsorbed due to the storing time under air. Nevertheless, the stability of the network was not influenced by the transfer process.

3.2 STM tip preparation[†]

A crucial factor for STM image quality is without any doubt the characteristics of the used tip. The origin of atomic resolution is facilitated by the shape of the tip, where close to 100% of the tunneling current is transported via the foremost apex atom.[18] Ideally, the tip is terminated with one single atom at the apex, free of contamination, and should have a small aspect ratio to reduce possible mechanical vibration. Of course, the apex of the tip should have atomically stable configuration during the raster-scanning process.

There exists a variety of approaches to form sharp metal tips that could be roughly classified as mechanical production of tips and others classified as physicochemical fabrication processes. Although several theoretical and experimental works were focused on tip fabrication,[104–107] it still remains difficult to produce tips that meet the high requirements. Nevertheless, optimal strategies for the tip fabrication process increase the yield noticeably.[108] Tips used under ambient conditions are mostly made of an alloy of Platinum and Iridium, $e.g.$ $Pt_{0.9}Ir_{0.1}$. They are produced by mechanical procedures such as cutting the wire at a certain angle with diagonal pliers, machining, or fragmenting. These mechanical tip formation approaches are very simple and time-saving. They facilitate scans with submolecular or even atomic resolution, specifically for less corrugated surfaces. Unfortunately, the tunneling junction of mechanically produced tips is normally less stable than for physicochemically produced ones because of numerous tiny asperities at the tip apex that may cause artifacts.

Thus, for UHV conditions the material of choice is tungsten due to its increased stability and inherent mechanical properties. In addition, the simple processing of sharp tips by using mild chemicals renders tungsten also suitable as probe material. Yet, the typically fabrication procedures, $i.e.$ electrochemical etching, are more complex. The etching process for tungsten wires (diameter 0.5 mm) follows the overall reaction:

[†]published [103]

$$W(s) + 8\,OH^- \longrightarrow WO_4^{2-} + 4\,H_2O + 6\,e^-$$
$$6\,H_2O + 6\,e^- \longrightarrow 3\,H_2(g) + 6\,OH^-$$
$$W(s) + 2\,OH^- + 2\,H_2O \longrightarrow WO_4^{2-} + 3\,H_2(g) \quad E^0 = -1,43\,eV$$

The basic etching setup consist of two electrodes, namely the tip and the counter electrode, both placed in an electrolyte solution. Applying a voltage in the range of 2 V to 5 V between these two electrodes leads to a dissolution of the tungsten wire. The sharp shape of the tip is caused by the fact that capillary forces are forming a meniscus of solution around the immersed tip wire that influences the etching rate. So, the etching rate at the bottom of the meniscus is a lot faster than at the top of the meniscus that leads to a tapering of the wire.[109] The quality of STM tips can be optimized by the choice and concentration of the electrolyte (KOH, NaOH) and with that its ion activity,[106] the applied voltage (range, AC/DC), and the used material, *i.e.* its purity. The finishing of the pre-fabricated tips is performed by means of a zone electro-polishing device under optical control in a light microscope.[110] Beyond that, it is worthwhile to note that tungsten is prone to oxidation under ambient conditions.[106,107] In order to remove the tungsten-oxide layer and etching remnants to achieve a stable tunneling junction, various procedures are proposed in literature such as ion-sputtering,[105,111,112] annealing by electron bombardment,[113,114] dipping into hydrofluoric acid (HF),[115] and self-sputtering in a noble gas environment.[116]

Before using the tips under ultra-high vacuum conditions, they have to undergo an after-treatment consisting of an *in situ* ion-sputtering. Subsequent thermal annealing is strongly recommended in order to remove tungsten oxides and etching residues. For this reason a versatile ion-sputtering and electron-beam annealing device (see Figure 3.3) was developed.

The basic components of the UHV compatible device for STM tip post-preparation are the axially arranged filament (tungsten wire, diameter 0.2 mm, 13 coils, grounded on the flange on the vacuum side), ring (tungsten wire, wire

Experimental Details and Materials

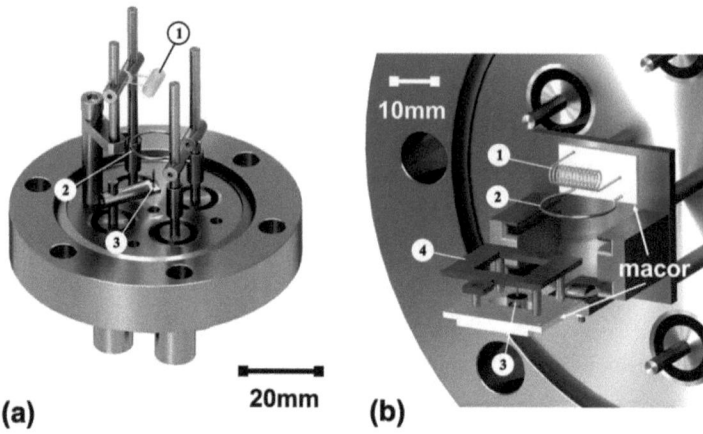

Figure 3.3: Scheme of the combined ion-sputtering and electron-beam annealing device: filament ①, ring electrode ② used as grid for ion-sputtering and used as a filament for electron-beam annealing, and tip holder ③. (b) Adaptation scheme of the proposed device presented in (a) for commercially available Omicron tip holders ④.[103]

diameter 0.2 mm, ring diameter 10 mm), and tip. This setup can easily be customized to a specific tip transfer and carrier system (see Figure 3.3 (b) for adaption to Omicron sample standard) and is therefore universally adaptable. Four high-voltage feedthroughs provide the wiring of the components. Due to adjustably mounted parts, the distances between tip and ring as well as between ring and filament are tunable in order to optimize the sputtering yield. To find optimal geometrical parameters, finite element simulation of the electrostatic potential and the electrostatic field was performed. The proposed device can be operated in two different modes, *i.e.* ion-sputtering and electron-beam annealing, just by changing the external wiring.

To operate the device in the ion-sputtering mode, noble gas (typically argon) has to be dosed into the UHV chamber via a leak-valve at a partial pressure of 1×10^{-5} mbar. Passing a DC current of 4.5 A through the filament yields thermal emission of electrons. A positive potential in the range of 800 V is applied to the ring assuring an acceleration of the electrons in this direction. During their acceleration to the ring, these electrons generate positively charged noble gas atoms by impact ionization. On their part, the argon ions are attracted by a negative potential of $-2000\,\text{V}$ onto the tip, resulting in a detectable sputter current. The ion current is positively correlated with both noble gas pressure and negative bias on the tip.

To operate the device in the electron-beam annealing mode, the ring is used as filament, *i.e.* as a source for thermal electrons. In contrast to the sputtering mode, a positive voltage in the order of $1500\,\text{V}$ is applied to the tip acting as a sink for electrons. After the sputtering, subsequent annealing is necessary to reduce defects which originate from the material removal during the ion-milling process. Here, the parameters, *i.e.* acceleration voltage, time of exposure, *etc.* have to be chosen very carefully to avoid blunting of the apex.[113]

In order to demonstrate the efficiency of the device, the prepared tips were characterized before and after sputtering as well as after annealing by scanning electron microscopy (SEM) in combination with spatially averaged energy dispersive X-ray (EDX) analysis. Directly after the electro-polishing procedure, the tip apex exhibits clearly visible contaminations (see Figure 3.4 (a) and (c)), which are due to etching remnants.

In addition – for approximately 50% of the probes – oxygen was detected by EDX confirming the presence of tungsten oxides which were also observed by other groups.[104,112] After the ion-sputtering process, the quality of the tips was assessed again by SEM and EDX characterization. Obviously, all visible contaminations were removed (see Figure 3.4 (b) and (d)) and the EDX spectra reveal no oxygen indicating the absence of tungsten oxides. Furthermore, the surface structure and possibly the tip shape were changed. Figure 3.5 presents SEM topographs of an electro-polished STM tip before and after annealing ($1500\,\text{V}$, $1.5\,\text{mA}$, $5\,\text{min}$). Apparently the shape of the annealed tip is changed because of partial melting of the tip and the onset of blunting. In this regard, it is important to mention that further studies of STM tips with different initial geometric shapes have shown that the resulting geometry is strongly related to the initial one. Therefore, it is hardly possible to provide general parameters for the annealing process. As a rule of thumb, we propose that for cone angles $\sim 25°$ no indication of blunting occurs after annealing times below $3\,\text{min}$ and

Experimental Details and Materials

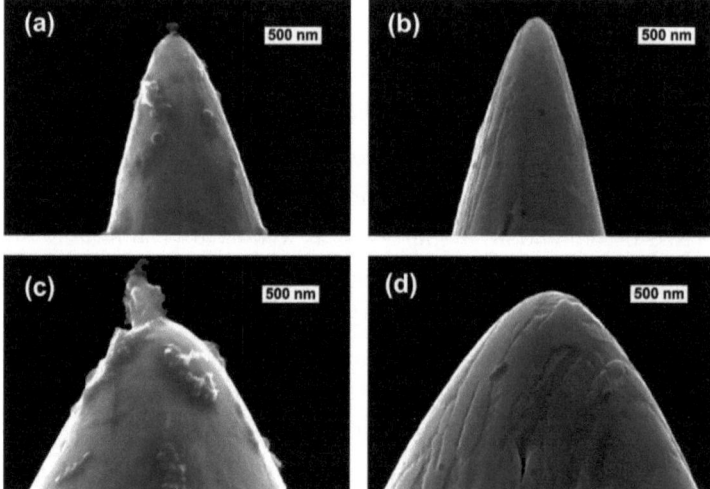

Figure 3.4: (a)/(c) present SEM micrographs of electrochemically etched tungsten tips directly after electrochemical etching and electropolishing without any further treatment. (b) illustrates the tip as shown in (a) after sputtering with 5 µA for 1 min, (d) the same tip as shown in (c) after sputtering with 10 µA for 5 min. Both examples clearly demonstrate that ion-sputtering in the proposed device is efficient for removing contaminations, but that also changes the surface structure, and possibly the tip shape. The detectable change of the outer shape of the tip shown in (d) is addressed to a tenfold increased ion-dose as compared to (b).[103]

currents of 1.5 mA at maximum acceleration of 1500 V. EDX revealed a further interesting result by detecting aluminum at the shank of the STM tip (see SEM images in Figure 3.5). The concerned parts were not electrochemically treated. We attribute this contamination to the wire drawing process.

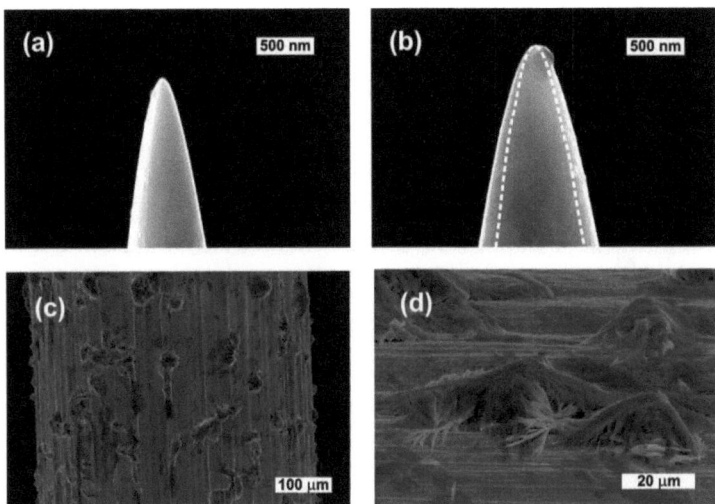

Figure 3.5: SEM micrographs of an electrochemically etched tungsten tip (a) before and (b) after electron-beam annealing (1.5 kV, 1.5 mA, *i.e.*, 2.25 W for 300 s). The dashed line in (b) illustrates the outer shape of the tip before annealing. The cone angle changes from 20° to 25°. (c) and (d) present SEM micrographs of the shank of a tungsten tip and a zoom in, respectively.[103]

EXPERIMENTAL DETAILS AND MATERIALS

3.3 Substrates, solvents, and molecules

Substrates In order to perform STM investigations, a solid, electrically conducting, and chemically stable sample is essential, which provides a versatile platform for steering and monitoring structures at the nanoscale.[28] The most commonly used substrate material to carry out STM measurements under ambient conditions is highly ordered pyrolytic graphite (HOPG), cleaved along the (0001) plane. Graphite surfaces are used because of their chemical inertness and the ease of preparing large and atomically flat areas. The (0001) direction is favored because of weak van der Waals interaction between sp^2 hybridized covalently bonded carbon layers within the stacked material. An additional consequence of this ABAB stacking is the increased lattice constant of 2.46 Å in STM images when just every other carbon atom is resolved. Figure 3.6 (b) shows an STM topograph that is also used as calibration standard for the lateral dimension. Since typical STM images have a size of the order 10 nm to 100 nm the domain size of the HOPG or the mosaic spread[‡] does not impair the experiment.

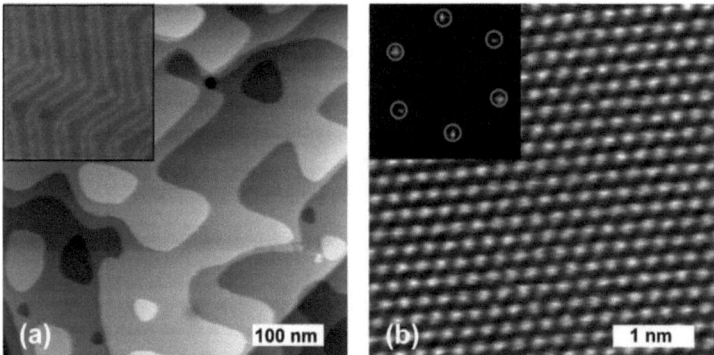

Figure 3.6: (a) STM topograph of a flame annealed Au(111) surface, the inset illustrates a 22 x $\sqrt{3}$ herringbone reconstruction caused by a contraction of the atomic spacing along the [1$\bar{1}$0] direction resulting in alternating hexagonal close packed (hcp) and face centered cubic (fcc) domains (b) STM topograph of atomically resolved graphite, inlay shows the 2D FFT image.

[‡]different quality grades caused by a variation of the c-axis orientation spread of the polycrystalline graphite, *i.e.* ZYA = 0.4° ± 0.1°, ZYB = 0.8° ± 0.2°

Also metals can be used as substrates but predominantly under UHV conditions due to their high reactivity. Under ambient conditions, the number of applicable metals is limited due to the lack of available preparation protocols and their tendency to oxidation and high vulnerability to undergo undesirable reactions with the air. Thus, the choice is often restricted to Au, which is of particular interest due to its bio-compatibility, non-toxicity, inertness, and available preparation methods under ambient conditions. Figure 3.6 (a) presents an STM topograph of flame annealed Au(111). In the context of the reaction on 2D surfaces the influence of the size of terraces, the abundance of step edges and kinks, the surface roughening, and the temperature dependent 2D adatom gas are decisive factors. Besides the material, the crystallographic orientation and consequently the related coordination number of the surface atoms[§] are of pivotal importance for governing surface mediated reactions. Therefore, every endeavor is made to prepare the desired crystallographic orientation (cf. section 3.4) which is *conditio sine qua non* either for studying surface properties at all or to serve as a template for epitaxial growth studies.

A very interesting type of new materials are graphene terminated substrates. To date, the one monolayer thick carbon sheets are available on various substrates, such as conducting Cu foil[117] or insulating SiO_2.[118] Graphene has outstanding properties, *e.g.* the longest known free mean path of electrons and highest current density at room temperature (10^6 times of copper) which arises from a combination of relativistic and quantum mechanical effects.[119] In addition, this modification of carbon has also excellent mechanical properties. Graphene is the thinnest imaginable, strongest, and stiffest known material but at the same time the most stretchable crystal (up to 20% elastic). Moreover, it is completely impermeable, *i.e.* even He atoms cannot penetrate and its thermal conductivity outperforms diamond. Figure 3.7 (a) shows a representative STM topograph of graphene terminated SiO_2, where graphene "wrinkles" were observed. They originate from the growth process of the graphene on the copper foil[117] and are not related to the transfer process onto the SiO_2 substrate. To characterize the

---

[§]Coordination numbers of a surface atom for a fcc crystal with (110), (100), or (111) orientation are 7, 8, and 9, respectively.

quality and the number of graphene layers, the prominent and rich features in Raman spectra, *i.e.* the G and 2D peaks, are examined (see Figure 3.7 (b)).

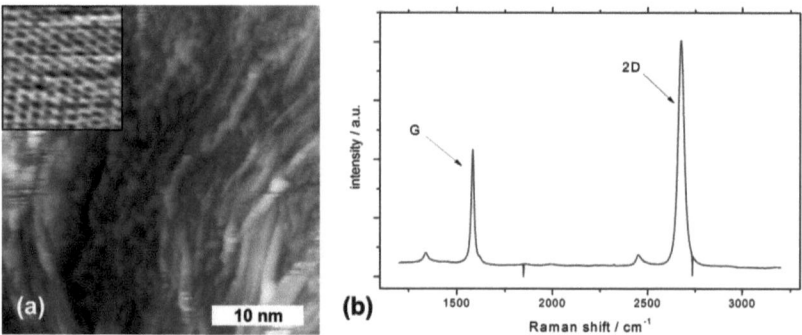

Figure 3.7: (a) Overview STM topograph of a graphene terminated SiO_2, where a wrinkled structure is obvious. The inlay 2 nm × 2 nm shows a zoom-in of the monolayer graphene structure. (b) Raman spectra of graphene terminated SiO_2. The line width of the 2D peak is 32 cm^{-1} and the I_G/I_{2D} ratio is approximately 0.5 both indicating a monolayer of graphene. STM tunneling conditions: $V_T = 0.280\,\text{V}$, $I_T = 87\,\text{pA}$.

A sharp line width $\sim 30\,\text{cm}^{-1}$ and a single Lorentzian profile of the 2D band at 2680 cm^{-1} provide an indication for a monolayer graphene. Furthermore, the ratio of the intensity of G to 2D peak I_G/I_{2D} provides a good correlation with the number of graphene layers.[120] For monolayer graphene, the I_G/I_{2D} ratio is approximately 0.5. On such substrates, even if the graphene was transferred to different materials, the graphene monolayer is not interrupted. Thus, it connects the adjacent terraces electrically over steps and kinks and compensates surface irregularities. In contrast to STM topographs of graphite, every atom within the graphene layer is visible, confirming a hexagonal lattice constant of 0.142 nm.

Moreover, ultra-thin films of insulating compounds, *e.g.* NaCl or RbI with a thickness of $\leq 10\,\text{Å}$ were also used as substrates for STM studies.[53] By means of such thin films, the interaction strength of the adsorbed molecules can be adjusted.

Solvents and molecules In the first instance, the insulating solvent is used to dissolve the targeted molecules. The solvent molecules interact among themselves, with the surface, and, dependent on their chemical characteristics, with the molecules under investigation. So the choice of the solvent can influence the final structure by offering different solvent-solute interaction.[121,122] Moreover, the solvent should have a low surface affinity in order to avoid full coverage with these molecules. Furthermore, attention should be foccused on the vapor pressure of the solvent, which influences the experimental conditions at the liquid-solid interface by varying the absolute concentration. In addition, the evaporation of the solvent can cause unstable tunneling conditions by thermal drift. As a rule of thumb, the *similia similibus solvuntur* rule is useful for choosing an appropriate solvent by taking the chemical functionalization of solvent and molecules into consideration. In this thesis, the investigated molecules under ambient conditions were predominantly dissolved in fatty acids (carboxylic acids with a long unbranched aliphatic tail) as presented in Table 3.1. It is worthwhile mentioning that the adsorption energy (on graphite) increases linearly with the fatty acid chain length.[123] Another important fact is the observation, that the concentration of dissolved molecules seems to vary with storage time, even for unsaturated solutions. These findings should be investigated in more details after the first preliminary UV-Vis adsorption spectroscopy experiments of coronene dissolved in nonanoic acid indicate a storage time dependency of concentration. Causal for these effects could be precipitation or clustering of the more likely dispersed than dissolved molecules.

In this thesis, organic compounds were used exclusively due to their relevance for further application in the scope of molecular electronics. Mainly, the molecular mass as well as dimension of molecules applied in the experiments was relatively small, *i.e.* 200 u to 700 u and 0.5 nm to 2.0 nm, respectively. These solid chemical compounds typically consist of an organic (aromatic) backbone, which facilitates in most cases a planar adsorption on the surface. These flat lying adsorption geometries favor lateral recognition on appropriate surfaces.[124] For the formation of laterally interlinked 2D patterns, a three-fold symmetry of the elementary building blocks is favored. In general, the final structure of the network and thereby their properties can be governed by the type of functionalization, *e.g.* carboxylic acid groups, thiols, or halogens. Precisely, the final patterns are stabilized either by chemical interactions, *i.e.* covalent bonds, or physical interactions, *e.g.* hydrogen bonds or van der Waals interaction of the elementary building blocks. Nowadays, a very wide range of flexibility in the structure is possible due to advances in chemistry.[34]

Structure	Details
	7A, $C_7H_{14}O_2$ Heptanoic acid, Oenanthic acid Molecular weight: $130.18\,\mathrm{g\,mol^{-1}}$, CAS number: 111-14-8 Vapor pressure: 1 hPa @ 20 °C
	9A, $C_9H_{18}O_2$ Nonanoic acid, 1-Octanecarboxylic acid Molecular weight: $158.23\,\mathrm{g\,mol^{-1}}$, CAS number: 112-05-0 Vapor pressure: 4 hPa @ 20 °C
	COR, $C_{24}H_{12}$, Coronene, [6]circulene Molecular weight: $300.36\,\mathrm{g\,mol^{-1}}$, CAS number: 191-07-1
	TMA, $C_9H_6O_6$ 1,3,5-Benzenetricarboxylic acid, Trimesic acid Molecular weight: $210.14\,\mathrm{g\,mol^{-1}}$, CAS number: 554-95-0 three-fold symmetry (D_3h)
	TPTC, $C_{22}H_{14}O_8$, p-terphenyl-3,5,3',5'-tetracarboxylic acid Molecular weight: $406.34\,\mathrm{g\,mol^{-1}}$ synthesized after known literature procedure[125]
	R = H, **TPB**, $C_{24}H_{18}$, 1,3,5-triphenylbenzene Molecular weight: $306.40\,\mathrm{g\,mol^{-1}}$, CAS number: 612-71-5
	R = COOH, **BTB**, $C_{27}H_{18}O_6$, 1,3,5-benzenetribenzoic acid, 1,3,5-tris(4-carboxyphenyl)benzene Molecular weight: $438.43\,\mathrm{g\,mol^{-1}}$, CAS number: 50446-44-1
	R = SH, **TMB**, $C_{24}H_{18}S_3$, 1,3,5-tris(4-mercaptophenyl)benzene Molecular weight: $402.59\,\mathrm{g\,mol^{-1}}$
	R = I, **TIPB**, $C_{24}H_{15}I_3$, 1,3,5-tris(4-iodophenyl)benzene Molecular weight: $684.09\,\mathrm{g\,mol^{-1}}$, CAS number: 151417-38-8
	R = Br, **TBPB**, $C_{24}H_{15}Br_3$, 1,3,5-tris(4-bromophenyl)benzene Molecular weight: $543.09\,\mathrm{g\,mol^{-1}}$, CAS number: 7511-49-1

Table 3.1: Chemical information of the regularly used solvents and investigated organic molecules. For further information please consult http://webbook.nist.gov/chemistry/.

3.4 Surface and sample preparation

STM experiments require a careful preparation of both extended atomically flat smooth surfaces and STM probes. In general, surface properties differ strongly from bulk material. Attributed to the broken bulk periodicity at the surface, unsaturated valences are accountable for relaxation, reconstruction, defects, or segregation.[126,127] As a consequence of this, surfaces often show different electronic, vibronic, optical, or magnetic properties compared to the bulk material properties.

At the liquid-solid interface, the sample preparation is straightforward. In particular, this is true for HOPG, where the topmost layers are just cleaved off using adhesive tape. Preparing Au(111) surfaces for ambient purposes requires either 12 seconds of O_2 plasma treatment whilst being heated to 100 °C[102] or thermal flame annealing using a Bunsen burner. To avoid contamination of the Au(111) on mica samples (purchased from Georg Albert, Heidelberg), the films were stored in pressurized nitrogen atmosphere. Moreover, there are electrochemical preparation protocols available to modify a metal surface by other metals with different coverage. Figure 3.8 (a) shows a Cu terminated gold surface with a mixed adlayer of 2/3 monolayer of Cu and 1/3 monolayer of sulfate utilizing underpotential deposition (UPD) in the vicinity of 0.15 V in steady-state cyclic voltammetry.[128–130] The copper atoms form a commensurate honeycomb lattice with a $(\sqrt{3} \times \sqrt{3})R30°$ symmetry while the sulfate ions are adsorbed in the centers. By selecting a voltage of approximately 0.50 V during the deposition process, a pseudomorphic monolayer of Cu(1 × 1) on Au(111) is attainable.[131] Depending on the applied potential, various surface coverages of Cu can be achieved.

To increase the number of available substrates having different properties such as lattice constant and corrugation, also iodine terminated gold substrates were prepared.[132] Besides an electrochemical way to create this modification, the immersion of a pre-annealed Au(111) film in a 3 mM potassium iodine solution for 180 s and subsequently rinsing with pure water is sufficient (see Figure 3.8 (b)). Depending on the exposure time and the concentration, different densely

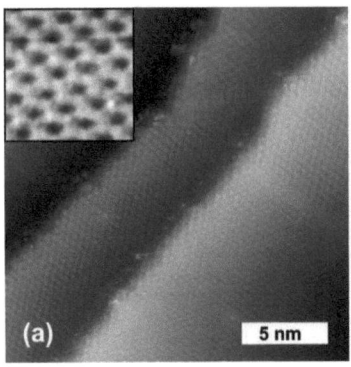 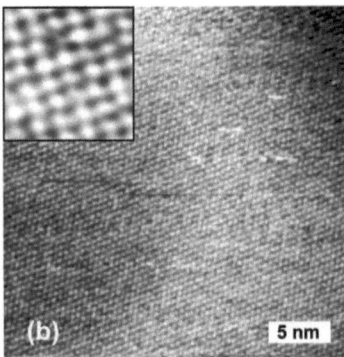

Figure 3.8: (a) STM topographs of an underpotential deposition of Cu, presenting an adlayer consisting of 2/3 monolayer of Cu on a Au(111) surface. The inlay 2 nm × 2 nm shows the commensurate honeycomb lattice of the Cu atoms with a $(\sqrt{3} \times \sqrt{3})R30°$ symmetry. The sulfate ions in the centers are not visible. (b) presents an STM topograph of an iodine terminated Au(111) film. The inlay 2 nm × 2 nm illustrates a zoom-in of the structure.

packed monolayers can be obtained.[132] The interaction strength of organic adsorbates with iodine terminated gold substrates compared to bare Au(111) surfaces is weaker. In general, the coating of metal surfaces with a monolayer of a different elements modifies the surface properties with respect to the bare bulk material.

After finishing the sample preparation, a droplet of solution ∼ 2.5 µL, consisting of organic solute molecules in the solvent, was immediately applied onto the surface. Subsequently, the STM tip was immersed into the liquid and brought to tunneling distance.

On the contrary, the sample preparation under UHV conditions requires major efforts. This is described for metals such as Au, Cu, and Ag in the following.

Sputtering Before using a crystal or thin metal film after exposure to air for the first time or starting a new experiment with a previously used crystal, preparatory work has to be done. It is imperative to remove the residuals from storage or former experiments and to optimize the surface quality, *e.g.* by removing scratches from polishing and reducing surface roughness. Thus, cycles of sputtering and subsequent annealing are essential.[133] For cleaning, bombarding the surface with noble gas ions (e.g. Ar^+, Ne^+) and subsequent annealing is the most commonly employed method. Typical ion energies are in the range of 500 eV to 1000 eV, and a current density of 5 µA cm^{-2} is optimal for removing surface contaminants together with the topmost atomic layers. The optimal duration of bombarding (typically 20 min) is dependent on the current density, the chosen number of layers to be removed, and the kind of material.[51] In order to detach embedded and adsorbed noble gas atoms and to recover the clean crystalline surface, subsequent annealing has to be performed. Heating the crystal to temperatures below its thermal roughening temperature[134] results in terraces separated by mono-atomic steps across the surface.[133] For quality inspection, *i.e.* to confirm cleanliness and ordering of the surface, low-energy electron diffraction analysis can by utilized.

Knudsen cell – Evaporating molecules To deposit molecules onto surfaces under vacuum conditions, various different techniques are available such as electro-spray deposition,[19,135,136] pulse injection,[137] and organic molecular beam epitaxy (OMBE).[138] Albeit OMBE carries the disadvantages that large fragile molecules can either be fragmented during the evaporation or some type of molecules are already reacting within the crucible, it is the most frequently used technique for sublimating molecules with sufficiently low vapor pressure. Despite these restrictions to evaporable components, large extended 2D networks on surfaces can be achieved by depositing one or more different types of molecular building blocks and subsequent activation of a bottom-up assembly mechanism.[139] To fulfill the requirements for this approach, at least two independently addressable evaporation cells are needed.

Hereafter, a double Knudsen cell evaporator based on the setup of Gutzler *et al.* is presented (see Figure 3.9).[140] Besides the sublimation of molecules, this design also enables to determine the evaporation rate as a function of temperature with the integrated quartz crystal microbalance. Based on these results, an approximation of the sublimation enthalpy of the molecules is possible. The power supply for the resistive heater (0.3 mm tantalum wire with defined coil density) is realized by feedthroughs. Type K thermocouples guarantee the independent temperature control of each molybdenum crucible and serve as control variable for the feedback system (Eurotherm 2416). Each cell is connected *via* separated feedtroughs to avoid cross talk between the different crucibles. A grounded BNC feedthrough connects the quartz crystal microbalance, which is integrated in the shutter, with the control electronics. The setup can be operated in a temperature range between 50 °C and 500 °C whereby each source can be addressed independently.

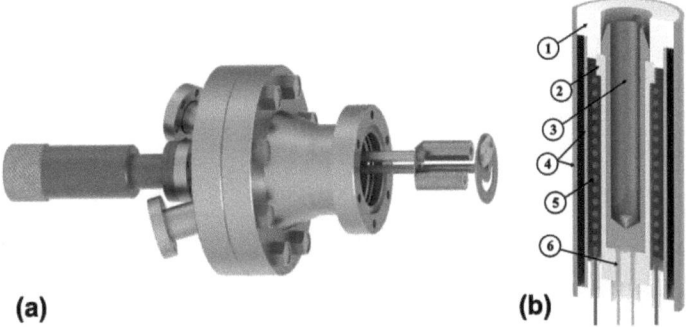

Figure 3.9: (a) Scheme of the enhanced double Knudsen cell based on the setup of Gutzler *et al.* The setup is realized on a DN 63 CF flange, but reduced to a DN 40 CF flange. (b) Detailed view on a revised single Knudsen cell. By means of a fixing screw touching the bevel plane of the crucible the bottom of it is pressed against the ball of the thermocouple. Thus a stable mechanical contact is established. The major assembly parts are a macor holder ①, a macor housing ②, a molybdenum crucible ③, a shielding ④, a tantalum wire for resistance heating ⑤, and a type K thermocouple ⑥.

It is strongly recommended to degas the compound in the crucibles before beginning an experiment because otherwise impurities of the powders could be deposited on the surface. Although molecules which are thermally evaporated from this type of effusion cell are distributed according to Knudsen's cosine law,[141]¶ the coverage of the area scanned by the STM is homogeneous due to the small field of view of the STM.

¶angular distribution of the number of molecules per unit area: $N = N_0 \cos^4 \alpha$, N_0 maximum number per unit area for emission angle $\alpha = 0$, perpendicular beam.

Experimental Details and Materials

In order to determine a suitable evaporation temperature for newly synthesized molecules, the change of mass and the related frequency shift (cf. Equation 3.1) as a function of time, i.e. $\Delta m(\Delta t)$, is recorded with the integrated quartz crystal microbalance while maintaining a constant temperature T at the crucible. Assuming that all or a constant fraction of the sublimated molecules stick on the quartz crystal, the total mass of the system increases, which causes a downshift of the resonance frequency of the quartz. Sauerbrey derived the following equation:[142]

$$\Delta m = -\frac{A v_{trans}\, \rho_Q}{2 f_0^2} \Delta f \tag{3.1}$$

where f_0 is the nominal eigenfrequency of the standard type of microbalance crystal (6 MHz), A the area of the quartz crystal, and ρ_Q the density of the crystal ($2.65\,\frac{g}{cm^3}$). The propagation velocity v_{trans} of an elastic transversal wave in direction perpendicular to the surface, i.e. the $01\bar{1}$ plane, is $3340\,\frac{m}{s}$.

Hence, the number of molecules sublimated per unit time is correlated with the impingement rate.[143] Assuming a constant temperature, a long mean free path of the molecules, and a constant sticking coefficient yields

$$\frac{\Delta m}{\Delta t} = \frac{1}{4}\frac{A}{RT}\sqrt{\frac{8RTM}{\pi}}\, p(T) \tag{3.2}$$

where R equals the ideal gas constant, M is the molar mass, and $p(T)$ is the saturation vapor pressure. Equation 3.2 reveals a simple proportionality between deposition rate $\frac{\Delta m}{\Delta t}$ and vapor pressure $p(T)$. To estimate the sublimation enthalpy ΔH_{sub} a substitution of the vapor pressure p with the deposition rate or the frequency shift ($\frac{\Delta f}{\Delta t}$) in the Clausius-Clapeyron-equation $\ln p \propto -\Delta H_{sub}/k_B T$ results in

$$\ln \frac{\Delta f}{\Delta t} = -\frac{\Delta H_{sub}}{k_B T} + C \tag{3.3}$$

where k_B is the Boltzmann constant, and C is an arbitrary constant which does not have to be evaluated. A plot of $\ln \frac{\Delta f}{\Delta t}$ versus reciprocal crucible temperature $\frac{1}{T}$ results in a straight line whose slope is $-\Delta H_{sub}/k_B$. Consequently, the

SURFACE AND SAMPLE PREPARATION

sublimation enthalpy per molecule can be obtained by subtracting out the Boltzmann constant. Alternatively, the sublimation enthalpy value per mol can also be extracted from the slope, as $R = k_B N_A$.

The above-mentioned strategy is applied to determine the sublimation enthalpy of a series of molecules composed of the same organic backbone, *i.e.* four phenyl-rings. They just differ in their functionalization. The results for 1,3,5 Tripenylbenzene (TPB) are exemplarily demonstrated in Figure 3.10.

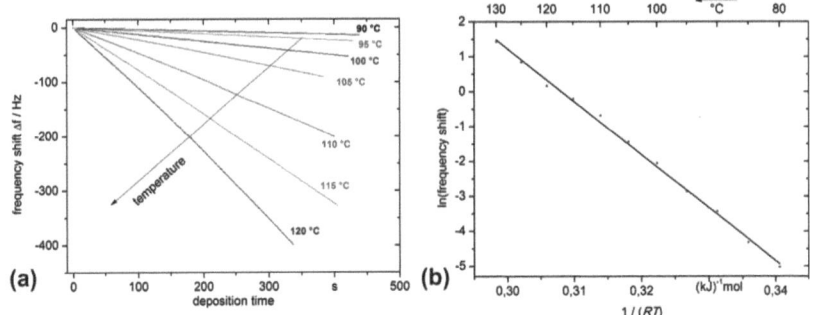

Figure 3.10: (a) The graph displays the frequency shift as a function of deposition time at different evaporation temperatures, *i.e.* from 90 °C to 120 °C, for 1,3,5 Tripenylbenzene (TPB). (b) An Arrhenius plot allows the determination of the sublimation enthalpy. Herein, the slope of different function in (a) is plotted versus the related inverse evaporation temperature times gas constant $\frac{1}{RT}$.

Figure 3.10 (a) shows the frequency shift as a function of the deposition time at different evaporation temperatures. The sublimation enthalpy ΔH_{sub} was determined from the slope of an Arrhenius plot of the frequency shift versus $\frac{1}{RT}$ to $(151.9 \pm 7.0)\,\text{kJ}\,\text{mol}^{-1}$. A comparison with values taken from literature for this temperature region (mean value $149\,\text{kJ}\,\text{mol}^{-1}$)[144] is in perfect agreement with the obtained results, which renders the approach suitable for the determination of sublimation enthalpies. Besides the sublimation enthalpy (see Figure 3.10 (b)) a temperature value for sublimation can be derived. For all practical purposes the most reasonable sublimation temperature can be taken, when the slope of the frequency shift is $3\,\text{Hz}\,\text{s}^{-1}$.

Using the same method, the sublimation enthalpy for 1,3,5-benzenetribenzoicacid (BTB) was determined to $(191 \pm 7)\,\text{kJ}\,\text{mol}^{-1}$ and for 1,3,5-tris(4-

mercaptophenyl)benzene (TMB) to $(151 \pm 6)\,\text{kJ}\,\text{mol}^{-1}$. Further, Gutzler *et al.* determined the sublimation enthalpy for 1,3,5-tris(4-bromophenyl)benzene to $(177 \pm 3)\,\text{kJ}\,\text{mol}^{-1}$.[140] The single contribution of each functional groups to the overall sublimation enthalpy of the entire molecule is dependent on the vicinity of the functional groups and the organic backbone. Thus, it is impossible to predict sublimation enthalpies just by simple arithmetic.

CHAPTER 4

Interactions at the nanoscale – from building blocks to 2D networks

This chapter deals with the fundamental interactions and processes at the nanoscale. Section 4.1 introduces the basic concepts of Gibbs free energy and self-assembly and also elucidates the influences of external parameters to these processes. In addition, transition state theory is introduced to describe covalent coupling through radical recombination. Since molecule-molecule interactions play an important role for the formation of 2D networks, they are described in section 4.2. Finally, section 4.3 sheds light on the interactions of adsorbed building blocks with the substrate, namely physisorption and chemisorption as well as surface mediated reactions.

4.1 Basic principles – processes at the nanoscale

Understanding the fundamental interactions of the involved building blocks among each other and with the substrate is a key challenge in 2D crystal engineering on surfaces. It is worth noting that the confinement to two dimensions already eliminates several degrees of translational, rotational, and vibrational freedom of the tectons.[23] Nevertheless, some of these 2D pattern are supposed to serve as initial templates to grow well defined 3D structures.

In 1828 Friedrich Wöhler laid the foundation for supramolecular chemistry by the synthesis of urea. This was the first time, a chemical compound was artificially synthesized from inorganic educts. Until now, the field of supramolecular chemistry has developed fast and brought up powerful methods for the construction of complex molecular structures by forming or breaking of non-covalent bonds between atoms in a controlled and depreciated manner.[145] In the toolbox of molecular chemistry, self-assembly* is of particular interest. Whitesides described this phenomenon as spontaneous formation of molecules into stable, well-ordered structures with non-covalent interactions between the building blocks.[146] The spontaneous process of self-assembly enables the formation of larger units out of small buildings blocks consisting of individual functional components.[139] The information for self-assembly is encoded in the involved components, their shape, topology, sequence of deposition, and specific surface properties. This reversible[147] process also occurs in several examples in nature (*e.g.* DNA double helix) and can be used as a blueprint for the adaptation of "artificial" materials. Prerequisite for molecular self-assembly are predefined tectons which only weakly interact and facilitate bond formation and bond breaking until a stable state of equilibrium is reached. However, the term self-organization is reserved for molecular systems that forms structures far from thermodynamic equilibrium.[27] In general, these processes occur in three dimensions but are restricted to surfaces in the following. Self-organization and self-assembly can be controlled to some degree by the ratio between molecular flux (**F**) and surface diffusivity (**D**). For high flux and low diffusivity, the molecules are trapped in a diffusion-limited state (see Figure 4.1 (a)). These kinetically stabilized structures are not settled in the global thermodynamic minimum. In contrast, for high diffusivity and low flux, the molecules are allowed to move freely on the surface ending up in the thermodynamically favored equilibrium structure.[34] Nevertheless, kinetically stabilized phases in an energetically meta-stable phase can be transformed within the time of the experiment into thermodynamically comparably stable polymorphs.[148] In general, thermodynamically controlled

---

*The term is also often used in biological systems, in which complementary units find each other and form a stable aggregate.[28]

systems minimize Gibbs free energy G. The change of Gibbs free energy ΔG is the appropriate measure

$$\Delta G = \Delta H - T\Delta S \tag{4.1}$$

where ΔH is the change in enthalpy, T the temperature, and ΔS the change of entropy of the system at constant temperature (isothermal) and pressure (isobaric). The overall entropy is composed of contributions from translational, rotational, vibrational, and conformational entropy. For liquid-solid environments, solvation and adsorption enthalpy are the main contributions to the enthalpy term of Gibbs' free energy equation. For $\Delta G < 0$ (exergonic), a spontaneous reaction is favored without the need of any external stimulus, for $\Delta G = 0$ neither the forward nor the reverse reaction is preferred because the system has reached its equilibrium state, and for $\Delta G > 0$ (endergonic) the reaction is disfavored and external stimulus is required. In the case of self-assembly ($\Delta G < 0$), the degree of order increases accompanied by a decrease of internal entropy. This contribution to the Gibbs free energy has to be at least compensated by a respective gain in enthalpy.

Without any doubt, the process of self-assembly is a balancing act between molecule-molecule (see section 4.2, E_{inter}) and molecule-substrate (see section 4.3, E_{ads}) interactions[149] and carries the potential for engineering advanced structures with a high degree of complexity. Figure 4.1 illustrates possible processes on a surface while Table 4.1 gives an instructive overview of the different interaction types in terms of energy range, typical bond lengths, and characteristics.

At room temperature, the typical adsorption energy per molecule E_{ads} should exceed 1 eV to prevent adsorbed molecules from desorption.[28] After the tectons are adsorbed on the surface – usually at specific sites due to their functional groups – they can diffuse between these different sites on the surface. This process obeys to a rate equation $\Gamma_{\text{diff}} = \nu_{\text{diff}} \exp\left[-\frac{E_{\text{diff}}}{k_{\text{B}}T}\right]$, where Γ_{diff} is the hopping rate, E_{diff} is the diffusion energy barrier, and ν_{diff} a prefactor. In the case of large adsorbed molecules under ambient conditions, a prefactor ν_{diff} in the range of $10^{10}\,\text{s}^{-1}$ to $10^{14}\,\text{s}^{-1}$ and a diffusion energy E_{diff} of 0.5 eV are reasonable assumptions. In order to "hop" on top of the surface, these adsorbed molecules

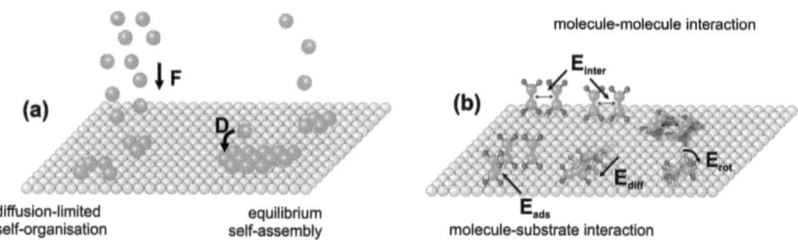

Figure 4.1: (a) Scheme illustrating molecular self-organization and self-assembly processes. Depending on the ratio of molecular flux (F) and diffusivity (D), either diffusion-limited self-organized structures (left) or self-assembled (right) structures are formed.[27] (b) Sketch of possible interaction at the nanoscale. The lateral interactions E_{inter} can be caused by van der Waals interactions, hydrogen bonding, electrostatic ionic interaction, dipole-dipole interactions, or metal-ligand interactions after the tectons are adsorbed onto the surface. Besides the adsorption energy E_{ads}, diffusion E_{diff} and rotational E_{rot} motion of the tectons are crucial processes occurring on the interface.

have to overcome the energy barrier, which is either determined by the particular adsorption-configuration of the system[150] or governed by anisotropic surface effects.[151] In analogy to lateral transport phenomena, tectons also need an activation energy to facilitate rotation. This process also obeys a rate equation, where similar barrier heights are expected for the rotation energy E_r as known for the diffusion energy barrier E_{diff} in most systems.

The mobility of the molecules can be changed by varying the substrate temperature. The applied thermal energy is transferred into kinetic energy E_{kin} of the molecules, whereby the diffusion barrier of the surface can be overcome.[34] Nevertheless, the kinetic energy of the molecules has to stay below the adsorption energy, otherwise this would cause desorption of the molecule from the surface. Moreover, for self-assembled structures, the bond strength of molecule-molecule interactions should be sufficiently weak in order to allow the molecular entities to optimize the structure and eventually reach an equilibrium with a global minimum of energy. On the other hand, the molecule-molecule interaction must be sufficiently strong in order to form stable structures. To obtain molecular self-assembly on surfaces, the following conditions have to be fulfilled: $E_\text{ads} > E_\text{inter} \geq E_\text{kin} > E_\text{diff}$. As a consequence of the weak interactions between molecules, reorganization and self-repair can easily be achieved.

	Energy range	Distance	Character
Van der Waals	$E_\text{vdw} \approx 0.02\,\text{eV}$ to $0.1\,\text{eV}$	$< 10\,\text{Å}$	Non-selective
Hydrogen bonding	$E_\text{h} \approx 0.05\,\text{eV}$ to $0.7\,\text{eV}$	$\approx 1.5\,\text{Å}$ to $3.5\,\text{Å}$	Selective, directed
Electrostatic ionic	$E_\text{el} \approx 0.05\,\text{eV}$ to $2.5\,\text{eV}$	$\approx 1.5\,\text{Å}$ to $2.5\,\text{Å}$	Non-selective
Dipole-dipole	$E_\text{dd} \approx 0.05\,\text{eV}$ to $2.5\,\text{eV}$	$\approx 1.5\,\text{Å}$ to $2.5\,\text{Å}$	Directional
Metal-ligand interactions	$E_\text{ml} \approx 0.5\,\text{eV}$ to $2.0\,\text{eV}$	$\approx 1.5\,\text{Å}$ to $2.5\,\text{Å}$	Selective, directional
Adsorption	$E_\text{ads} \approx 0.5\,\text{eV}$ to $10\,\text{eV}$	$\approx 1.5\,\text{Å}$ to $3\,\text{Å}$	Directional, site selective
Surface diffusion	$E_\text{diff} \approx 0.05\,\text{eV}$ to $3\,\text{eV}$	$\approx 2.5\,\text{Å}$ to $4\,\text{Å}$	1D / 2D
Rotational motion	$E_\text{rot} \approx \dim(E_\text{m})$	s	2D
Substrate mediated	$E_\text{sm} \approx 0.001\,\text{eV}$ to $0.1\,\text{eV}$	Nanometer range	Oscillatory
Reconstruction mediated	$E_\text{rm} \approx 1\,\text{eV}$	Short	Covalent

Table 4.1: Overview of basic interactions and processes at the nanoscale, with associated energy and typical distances, adapted from [28,34].

Transition state theory (TST) – Ullmann reaction For a qualitative description of chemical reactions, in particular for the Cu-mediated Ullmann coupling[†] of two aryl halides,[152,153] transition state theory is appropriate.[154,155] Herein, a hypothetical energy state that exists during a chemical reaction between reactants and products is described. The transition state is primarily composed of unstable regions of the energy landscape which lead to a quick stabilization into the reactant or product basins. In Ullmann reaction the intermediate state involves most likely a bond of the aryl to a Cu^{+I}.[156,157] However, also more complex mechanisms such as single electron transfer were discussed.[158] At this intermediate state, reactants are combined in order to form species called the activated complexes, which are in quasi-equilibrium with the reactants. A reaction profile diagram (see Figure 4.2) allows to plot the respective energies, i.e. the standard enthalpy of activation $\Delta^\ddagger H$, the standard entropy of activation $\Delta^\ddagger S$, and the standard Gibbs energy of activation $\Delta^\ddagger G$, in dependence of a reaction coordinate under the assumption of a known rate constant.

In addition, transition state theory can be utilized to predict, whether or not a reaction will take place, depending on the population of the activated complex.

[†]The Ullmann reaction is named after Fritz Ullmann. It includes a copper-catalyzed nucleophilic aromatic substitution between various nucleophiles (e.g. substituted phenoxides) with aryl halides.

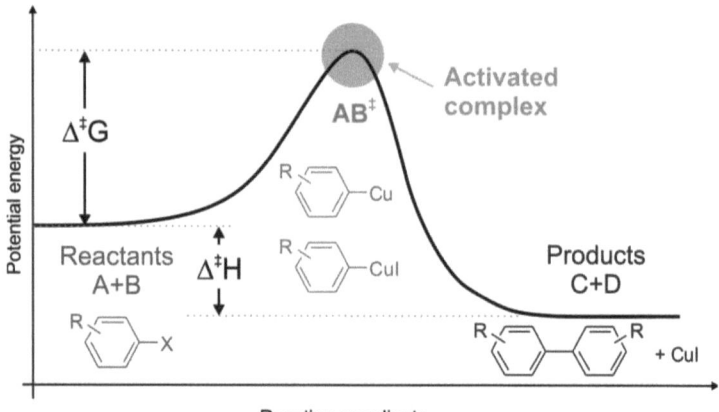

Figure 4.2: Reaction profile of a hypothetical exothermic reaction of reactants (A + B) to products (C + D). On the axis of the abscissas, the progress of the reaction is plotted using the example of an Ullmann reaction. The activated complex (transition state AB‡) is marked with a light gray shadowed area.

The rate at which the activated complex breaks apart, and the way it breaks apart (to reconstitute the reactants or to form a new complex, *i.e.* the products) can also be described with this theory. The rate constant k of an elementary reaction between reactants and activated complexes, which are assumed to be in a special kind of equilibrium, is given by the Eyring equation:

$$k = \kappa \frac{k_\text{B} T}{h} \exp\left[-\frac{\Delta^\ddagger G}{RT}\right] \quad (4.2)$$

where k is the reaction rate constant, κ the transmission coefficient that is often assumed to have a value of unity, k_B is the Boltzmann constant, R the gas constant, h Planck's constant, and $\Delta^\ddagger G$ the Gibbs energy of activation. Since $\Delta G = \Delta H - T\Delta S$, Equation 4.2 can be expressed as:

$$k = \kappa \frac{k_\text{B} T}{h} \exp\left[\frac{\Delta^\ddagger S}{R}\right] \exp\left[-\frac{\Delta^\ddagger H}{RT}\right] \quad (4.3)$$

where $\Delta^\ddagger S$ is the entropy of activation, which is the standard molar change of entropy when the activated complex is formed from reactants, and $\Delta^\ddagger H$ is the enthalpy of activation. Indeed, the energy of activation and the enthalpy of activation $\Delta^{\ddagger}H$ are not quite the same, depending on the type of reaction. In this context, it has to be mentioned that the fundamental equation of TST resembles the Arrhenius rate equation $k = A\exp\left[\frac{-E}{RT}\right]$ and nevertheless offers an interpretation of A, the pre-exponential factor, and E, the activation energy.

Besides the molecular structure addressed above, external conditions such as solvent, concentration, and temperature have an effect on the formation of 2D patterns because they change the kinetics and thermodynamics of the processes.

Solvent effects A major role in the formation of two dimensional networks at the liquid-solid interface is taken by the used solvent. It is capable of determining the final structure by the interaction between adsorbate and solvent. Due to great differences in the properties of liquids (polarity, viscosity and therewith related diffusivity of the molecules, and vapor pressure), the dissolved molecules may assemble in various structures at the liquid-solid interface.

Figure 4.3, for instance, illustrates different self-assembled phases of saturated 1,3,5-cynabiphenyl-benzol (CBPB) solution. The STM image Figure 4.3 (a) was acquired in heptanoic acid whereas (b) and (c) were obtained in nonanoic acid. The latter STM topograph was acquired on a Au(111) surface in contrast to the others that were obtained on a HOPG surface. For heptanoic acid (7A), the CBPB molecules form double rows of densely packed molecules. As clearly visible in Figure 4.3 (a), adjacent double rows are separated by the same distance. A detailed view on the rows suggests that hydrogen bonds between the cyano nitrogen atoms of the nitrile group and hydrogen atoms of the phenyl rings stabilize the entire structure. This is similar to 3-phenyl-propynenitrile on the Cu(111) surface, as reported by Luo et al.[159] On the other hand, using nonanoic acid (9A) as solvent while keeping all the other parameters constant yields a more regular spacing between the rows of CBPB molecules in accordance with some variation of the contrast. This phenomenon is very likely based on different adsorption sites of the CBPB molecules on the basal plane of the graphite. The different phases are most likely a consequence of co-adsorbed solvent molecules, that influence the spacing between the CBPB molecules and thus the strength of intermolecular interaction, e.g. van der Waals interaction and hydrogen bonds. As a consequence, the CBPB molecules are locked at different preferred adsorption sites due to the different geometric circumstances, e.g. length of the used fatty acid within the motifs. However, irregularly assembled structures were obtained when applying 1,3,5-cynabiphenyl-benzol dissolved in nonanoic acid (9A) on a Au(111) surface. In Figure 4.3 (c), the three-fold symmetric molecules are present and clearly discernible by their outer contour. This demonstrates that the choice of the substrate can suppress primary solvent effects by increasing the strength of substrate-molecule interaction in comparison to the strength of molecule-molecule interaction.

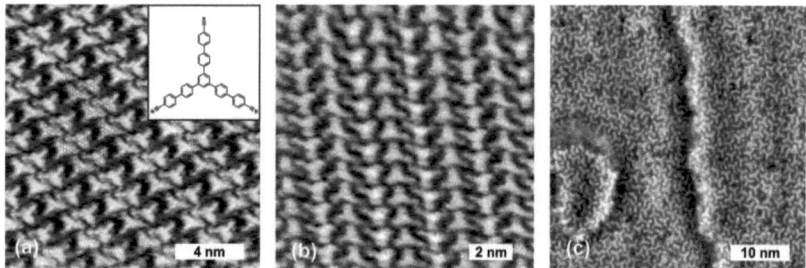

Figure 4.3: STM topographs of saturated 1,3,5-cynabiphenyl-benzol (CBPB) on HOPG dissolved in (a) heptanoic acid and in (b) nonanoic acid. (a) The image shows a typical obtained contrast, where a similar distance between adjacent rows was determined. The inset represents the chemical structure of CBPB. (b) STM topograph of the same system, while nonanoic acid was used as solvent. Here, a more regular spacing between the rows is clearly visible. (c) Adsorption of CBPB on Au(111) leads to irregular patterns, where the organic molecules are discernible by their three-fold symmetry. STM tunneling conditions: (a) $V_T = 0.35\,\mathrm{V}$, $I_T = 56.0\,\mathrm{pA}$, (b) $V_T = 1.43\,\mathrm{V}$, $I_T = 78.5\,\mathrm{pA}$, (c) $V_T = 0.96\,\mathrm{V}$, $I_T = 7\,\mathrm{pA}$.

Walch et al. presented a liquid-solid STM study, where the influence of the used fatty acid on the final structure of a hexagonal melamine network was investigated.[121] In these self-assembled networks, the fatty acid molecules were integrated in the structure. Each additional carbon atom in the tail of the solvent molecules causes an increase of 0.15 nm in the lattice constant.

Another aspect worth mentioning in this context is solvent induced polymorphism in supramolecular networks at the liquid-solid interface. Lackinger *et al.* reported two different structures of trimesic acid (TMA) on graphite, *i.e.* the flower structure and the chicken-wire structure.[100] Dependent on the used solvent and thus the length of the fatty acid, a certain TMA structure is favored. For short-chain fatty acids ($CH_3(CH_2)_n COOH$ with $n \in [2, 7]$) a flower structure was observed. However, for long-chain acids (n > 7) the chicken-wire structure was formed. Kampschulte *et al.* studied the behavior of an analog molecule, *i.e.* 1,3,5-benzenetribenzoicacid (BTB), where a polymorphism was evident as well.[101]

Concentration The final structure of 2D molecular networks at the interface is often dependent on the concentration of the building blocks in solution. By adjusting the concentration of the building blocks (dehydrobenzo[12]annulenes) in the liquid reservoir (1,2,4-trichlorobenzene), Lei *et al.* were able to generate different surface patterns in a controlled way.[160] The pores of these networks were stabilized by interdigitation of the alkoxy chains of the building blocks. The diameter of the pores reached a value of 5.4 nm which is by no means a fundamental limit and strongly dependent on the concentration. This dilution principle of surface self-assemblies can also be transferred to vacuum conditions. Here, the concentration dependence of adsorbed molecules on various surfaces is called "surface dilution principle".[161] For instance, reducing the surface coverage and with this the molecular packing density of trimesic acid on Au(111) leads to various structures with different interpore distances.[162]

Temperature A crucial factor not only for self-assembly is the temperature of the substrate, which primarily influences the mobility of adsorbed molecules and also provides access to controlled chemical reactions, *i.e.* catalysis.[32] Generally speaking, temperature is decisive in nearly every process governed by kinetic or thermodynamic factors on this length scale.[11,124,163] Occasionally, thermally activated processes such as diffusion and rotation of molecules have to be suppressed by cooling the sample down to low temperature in order to clarify the structures by means of STM. For controlling and varying the temperature, UHV conditions are appropriate. Under ambient conditions, the available range of temperature is restricted. Nevertheless, Gutzler *et al.* used temperature controlled liquid-solid STM experiments to reveal a reversible phase transition of 1,3,5-benzenetribenzoic acid from a nanoporous low-temperature (25 °C) phase to a more densely packed high-temperature (55 °C) phase.[164]

In general, there are several internal parameters such as geometry and functionalization of the molecular building blocks as well as external parameters such as used solvent, concentration, temperature, surface, orientation, and steric hindrance that govern the resulting self-assembled structures. The corresponding variety of possible combinations makes it hard to predict a structure, but also provides a great scope to create new advanced materials.

4.2 Interaction between molecules

The following section provides a short overview on the interactions between molecules adsorbed on surfaces, which are relevant for this thesis. In Table 4.1 the general characteristics of these forces, their energy range, and their typical bond distances are specified.

$\pi - \pi$ interactions $\pi - \pi$ interactions cause an attractive force due to electrostatic interaction. This type of interaction is of pivotal importance, because the molecules investigated here are mostly planarly adsorbed due to their conjugated aromatic systems. Hunter *et al.* proposed a simple model to describe favorable net $\pi - \pi$ interactions ($\pi - \pi$ stacking) for parallel oriented organic molecules, describing each molecule separated in a σ-framework and two π electron clouds. Thereby the $\pi - \sigma$ attraction between the molecules overcomes the repulsions of the electron clouds. This results in an attractive interaction.[165] Björk *et al.* calculated the adsorption energy of polycyclic aromatic hydrocarbons (PAH) physisorbed on graphene using van der Waals density functional (vdW-DF) theory.[166] Herein, the binding energy per carbon atom for the aromatic hydrocarbons was calculated and turned out as proportional to the ratio of hydrogen to carbon atoms. These calculated values are in good agreement with the results of temperature programmed desorption studies of the PAH. For instance, the adsorption behavior of coronene in a trimesic acid host network on graphite is presented in detail in section 5.1.

Van der Waals Interactions (vdW)   are probably the most prominent type of non-directional interactions.[167] Despite the fact that these interactions are often considered weak,[‡] they dominate the behavior of neutral systems.[169] VdW interactions originate from thermal and quantum charge fluctuations of neutral molecules and are therefore always present *in vitro* as well as *in vivo*. The main contribution to the overall force balance includes:

[‡]Nevertheless, Geckos facilitate vdW interaction to climp up smooth vertical surface or stick to virtually any surface. Therefore, the animal brings their superfine, flexible bristles (spatulae) in close distance to the surface and thus vdW interactions become effective.[168]

- the attractive or repulsive electrostatic interactions between permanent charges and between permanent multipoles in general (Keesom force)
- the attractive polarization force between a permanent multipole on one molecule with an induced multipole on another (Debye force)
- the attractive dispersion force between any pair of molecules, including non-polar atoms arising from the interactions of multipoles (London dispersion force).[170,171]

VdW interactions have anisotropic character, leading to a dependence on the relative orientation of the molecules. In order to describe the vdW interaction between two neutral particles as a function of distance the Lennard-Jones potential,[172] V_{vdW} is often used as an approximation. Herein, the repulsive portion is approximately proportional to r^{-12}§ and the attractive contribution scales with r^{-6}:

$$V_{vdW} = V_{repulsive} + V_{attractive} = 4\varepsilon \left\{ \left(\frac{\sigma}{r}\right)^{12} - \left(\frac{\sigma}{r}\right)^6 \right\} \quad (4.4)$$

where ε is the "depth" of the potential well at the equilibrium distance, σ a finite distance where the potential is zero, and r the distance between the particles. These parameters are accessible *via* quantum chemical calculations or can simply be fitted to reproduce experimental data. Moreover, the contribution of vdW forces to control the final structure of self-assembled layers and the competition of vdW interactions with other types of interactions such as hydrogen-bonds are well documented in literature.[173–175] For the visualization of molecules, the so-called van der Waals surface, an imaginary surface unifying the spherical surfaces (van der Waals radii) of the involved atoms is often used.

Hydrogen bonds or hydrogen bridges are omnipresent in chemistry and biology. Their nature is based on a complex combination of electrostatic interactions, polarization effects, covalency, and van der Waals interactions.[176] It is not surprising that also nature utilized these type of interaction to create the most

---
§Also a very common approximation is assuming an exponential repulsive part.

fundamental natural example of a molecular recognition system: the DNA. These non-covalent interactions between the DNA base pairs link the two complementary strands into a double helix. This type of highly directional and selective hydrogen bonds can also be utilized in nanotechnology to tailor and design 2D structures.[23] Steiner proposed the following definition for hydrogen bonds:

"An X-H $\cdots$ A interaction is called a hydrogen bond, if 1. it constitutes a local bond, and 2. X-H acts as proton donor to A."[177] The X-H group is called the (proton) donor and A is called the (proton) acceptor, respectively (see Figure 4.4 (a)). Typically, the strength of this interaction scales with the electronegativity of X and A. Depending on the internal structure of the molecules, *i.e.* the functional groups and the (organic) backbone, hydrogen bonds show a high variation in strength from very weak interactions ($CH_4 \cdots FCH_3$, $0.8\,kJ\,mol^{-1}$)[178] up to very strong interactions ($[F-H-F]^-$, $163\,kJ\,mol^{-1}$).[179] Besides their length dependence also the bond angle is variable rendering hydrogen bridges versatile for 2D crystal engineering. This allows topologically flexible interconnections in supramolecular architectures.[180]

C-H $\cdots$ O and O-H $\cdots$ O hydrogen bridges are of particular interest due to the fact that oxygen and hydrogen atoms are present in various organic molecules.[181] The emphasis is on carboxylic acid groups (COOH), which combine within a single functional group both hydrogen-bond donors (*via* the hydroxyl group $R-O-H$) as well as hydrogen-bond acceptors (*via* the carbonyl oxygen atom). In the case of cyclic double hydrogen bonds, resonance assisted hydrogen bonding (RAHB)[182,183] has to be taken into account. This special type of hydrogen bonds is stronger than equivalent single hydrogen bonds, whereas standard force fields such as the widely used MM3 fail to predict its energy and thus underestimate the strength of cyclic double hydrogen bonds. Such cooperative resonance effects are well known in molecular orbital theory, where the total energy of multiple bonds is larger than the sum of the individual components. Applying such tailored building blocks onto surfaces allows to form a wide range of 2D structures.[184–187] Common binding motifs (see Figure 4.4) for carboxylic acid groups are dimers, trimers, and catemeric motifs.[188] Figure 4.4 (b) - (c) presents

INTERACTIONS AT THE NANOSCALE

STM topographs of self-assembled structures, that are stabilized by hydrogen bonds. The STM topograph in Figure 4.4 (b) shows an image of the final structure of 4,4'-stilbenedicarboxylic acid (SDA) on the nonanoic acid-graphite interface. Herein, the two π-systems of the phenyl-rings, which are interconnected by an ethenyl unit, are clearly discernible. The one-dimensional rows within the large domains are stabilized by double hydrogen bonds, formed within the typical bonding distance of 0.25 nm.[189] Figure 4.4 (c) and (d) present STM topographs of 1,3,5-benzenetribenzoic acid (BTB) and 4-2-3,5-bis[2-(4-carboxyphenyl)-1-ethynyl]-2,4,6-trimethyl-phenyl-1-ethynyl benzoic acid, respectively. Both types of building blocks can be dissolved in nonanoic acid and applied on top of a graphite surface. The obtained honeycomb structures are stabilized by cyclic double hydrogen bonds between carboxylic groups in a self-complementary manner.

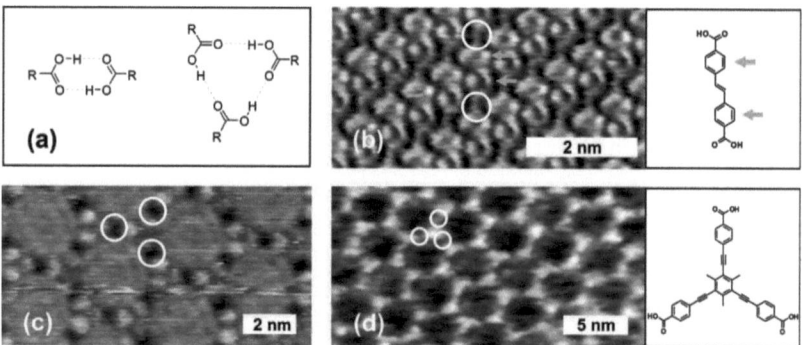

Figure 4.4: (a) Sketch of typical binding motifs formed between carboxylic acid, *i.e.* a cylic dimer (left) and a cyclic trimer (right). The STM topographs show different self-assembled structures at room temperature on graphite that are governed by hydrogen bonds, highlighted by white circles. The used solvent was nonanoic acid. STM topograph (b) presents a 4,4'-stilbenedicarboxylic acid (SDA) network, $V_T = 0.35\,\text{V}$, $I_T = 56\,\text{pA}$, (c) the self-assembled structure of BTB, $V_T = 0.48\,\text{V}$, $I_T = 47\,\text{pA}$, and (d) the hexagonal structure of 4-2-3,5-bis[2-(4-carboxyphenyl)-1-ethynyl]-2,4,6-trimethyl-phenyl-1-ethynyl benzoic acid, again a dimeric hydrogen bonded network, $V_T = 1.24\,\text{V}$, $I_T = 56\,\text{pA}$.

Halogen-halogen interactions cause directed intermolecular forces that are comparable in strength to hydrogen bonds.[190] This type of interaction that is not well known can be involved in crystal engineering, forming 1D, 2D, and 3D architectures,[191,192] in the construction of porous structures[193] as well as in biological application.[194] The nomenclature of halogen bonds is analog to the hydrogen bridges, just the hydrogen atom is substituted by the halogen atom leading to: D-X···A, where D is the donor, X is the halogen atom acting as Lewis acid, and A is the electron-rich species, *i.e.* the acceptor. Consequently, the type of interaction is a donor-acceptor-relationship where the strength follows the general trend: F < Cl < Br < I.[195] Bosh performed calculations of iodobenzene showing a non-spherical charge distribution leading to an attractive force between the positive potential at the cap opposing the C-I bond and a negative ring potential around the bond axis.[196] Depending on the organic backbone, the strength varies. For instance, the hybridization of the carbon atom next to the halogen influences the interaction strength: $sp^2 > sp > sp^3$.[197]

Metal-ligand bonds are highly specific directional interactions between a metal atom and an organic compound, *e.g.* carboxylic groups and alkoxides, that form robust entities in which the incorporated metal atoms provide specific functions (*e.g.* electronic, magnetic, and catalytic).[28] Materials based on these interactions allow for many possible applications in gas storage and gas separation, chemical sensing, as well as medical applications.[198–200] The thermodynamic stability of such complexes is increased by the number of ligands and the strength of the bonds. Moreover, an increasing number of ligands can lead to ligand repulsion, where the size of the metal atoms can become a limiting factor. For many complexes, a valence shell with 18 electrons is the most stable, as this is a restatement of the octet rule with additional 10 *d*-electrons.

Metal-ligand bonds have also been used not only to form 3D structures but also to form highly ordered metal-organic arrangements on surfaces.[201,202] Zacher *et al.* focused on surface chemistry of MOFs at the liquid-solid interface in their review.[203] Another example is the work of Schlickum *et al.* who formed honeycomb nanomeshes with tunable cavity sizes. They used ditopic dicarbonitrile-polyphenyl molecular linkers that were coordinated *via* cobalt (Co) on an Ag(111) surface.[204] These types of networks were extended over micrometers as a single domain and remained thermally stable up to 300 K. Thus, this type of network provides an opportunity for adsorption of guest molecules.

Covalent bonds are characterized by a balance of attractive and repulsive forces between atoms which are sharing one, two, or three electrons forming single, double, or triple bonds. The characteristics of the covalent bond is affected by the electronegativity of the involved atoms. Equal electronegativity between the binding partners leads to a homopolar covalent bond. In contrast, unequal conditions create a polar covalent bond. Molecular orbital theory is used to describe the bonding between atoms combining the wave-like characteristics of their atomic orbitals to form a molecular orbital. Typically, bonding molecular orbitals stabilize the newly formed molecule since less energy is associated as opposed to a system of unbound atoms. In contrast, anti-bonding molecular orbitals destabilize the system due to the higher energy. Normally, these bonds are distinguished by the symmetry and number of involved atomic orbitals. This means that σ-bonds have a head-on overlapping between the atomic orbitals, π-bonds form a molecular orbital by an overlap of two lobes of an appropriate atomic orbital, and a δ-bond is formed by an overlap of four lobes of one involved atomic orbital with four lobes of the second involved atomic orbitals. In the framework of this thesis, most of the investigated molecules are comprised by a combination of σ-bond along with π-bonds.

The most noted 2D covalently interlinked network is graphene, which was only discovered and fabricated in 2004.[81] Since then, great effort has been put into the synthesis of such 2D covalently bonded networks based on different tectons. The band gap of these graphene-like 2D structures is strongly related to the choice of the building block. In the last years, covalent organic networks were intensely investigated,[205,206] because of their outstanding mechanical and electronic properties and their huge potential for various applications. In 2007, Grill *et al.* reported a bottom-up construction strategy for sophisticated electronic circuits based on the covalent coupling of individual functionalized molecules for the first time.[207]

4.3 Interaction between molecules and surfaces

A fundamental understanding of interactions between adsorbates and metal surfaces as well as graphite is the prerequisite for understanding the complex processes on surfaces such as self-assembly, diffusion, growth of thin films, or catalytic reactions.[208] On surfaces, the periodicity of the crystal is breached by unsaturated bonds and thus the properties are changed with respect to the bulk. Moreover, the number of electrons associated with the organic molecules and the substrate varies extremely, which is a major difference to chemical bonds just among molecules.

Physisorption This type of exothermic and reversible interaction affects all atoms and molecules which get adsorbed on a surface. However, for reactive species and surfaces, this adsorption state may just be a precursor state before being converted in the final form of a chemical bond. Particularly, on less reactive surfaces such as graphite, the interaction between substrate and molecules is dominated by physisorption. Normally, the geometrical and electronic structure of adsorbent and adsorbates remains rather unmodified. Physisorption is strongly associated with van der Waals attraction. In contrast to vdW interaction between atoms, where the attractive interaction scales with r^{-6}, the attractive forces between adsorbent and adsorbates decays with r^{-3} implying a higher reach of the interactions. This reduction can be elucidated by the interaction between the charges outside the two dimensional surface and their images inside. Consequently, the attractive forces between adsorbate and surface are described as $E_{\text{att}} = -c_1 r^{-3}$, where c_1 is a product of dipole moments. One of theses dipole moments is attributed to the adsorbate and the other one to the image dipole scaling with the polarizability and the adsorbate dipole moment, respectively. However, at closer distances (electron clouds of the adsorbate start overlapping with the substrate) short range Pauli repulsion also has to be considered: $E_{\text{rep}} = c_2 \exp\left[-c_3 r\right]$ where c_2 and c_3 are constants. Therefore, the interaction potential has a steep increase at short distances. The overall interaction is given by the following expression:

$$E_{\text{vdW}} = E_{\text{rep}} + E_{\text{att}} = c_2 e^{-c_3 r} - \frac{c_1}{r^3} \tag{4.5}$$

Figure 4.5 shows a typical total potential of a physisorption process. The shape of the curve is characterized by a very shallow minimum of only a few meV which is located rather far away from the surface, typically more than 3 Å. At room temperature ($kT \sim 25\,\text{meV}$), a variety of physisorbed molecules are rather mobile and can diffuse on the surface while there is no interaction taking place among adsorbed adjacent molecules. Consequently, raising the temperature of surfaces decorated with physisorbed molecules can reduce the surface coverage of the species as a consequence of an increasing desorption rate. On the other hand, the retention time of physisorbed molecules on a surface kept at 100 K is in the order of seconds, whereas at room temperature it is just $10^{-8}\,\text{s}$.

Chemisorption originates from the formation of actual covalent or ionic bonds between adsorbed molecules and the surface, where the elementary step often involves an activation energy. This irreversible process of chemisorption, which is almost always spontaneous, as well as physisorption both require a negative change in the Gibbs free energy ΔG. For a negative change in enthalpy ΔH an exothermic reaction occurs. Since the adsorbed molecules are more ordered and consequently lose their freedom of movement, also the entropy ΔS change is negative. With this the requirements for a spontaneous reaction, *i.e.* $\Delta G < 0$ are fulfilled. However, after all molecules have occupied a certain adsorption site on the substrate, *i.e.* a mono-molecular layer of the species is formed, further chemisorption is blocked. These chemical adsorption processes necessarily lead to a strong change within the adsorbed molecules. Thus, strong interaction between adsorbate and substrate can lead to bond break within the adsorbed species. This bond weakening or dissociation is also the basis for chemical reactions such as heterogeneous catalysis (see next paragraph).

In contrast to physisorption, the enthalpy of the chemical adsorption, *i.e.* $40\,\text{kJ}\,\text{mol}^{-1}$ to $800\,\text{kJ}\,\text{mol}^{-1}$, is approximately one order of magnitude higher. This higher energy also leads to a shorter bond length of chemisorbed atoms and molecules. For instance, the bond length of a chemisorbed sulphur atom on a Ni(111) surface is 2.3 Å,[209] and for chemisorbed acetate on a Cu(111) surface the Cu-O bond distance amounts 1.91 Å,[210] respectively.¶ Figure 4.5 shows a general potential energy diagram with two intersecting curves, illustrating the process of molecular physisorption (curve 1) and dissociative chemisorption (curve 2).

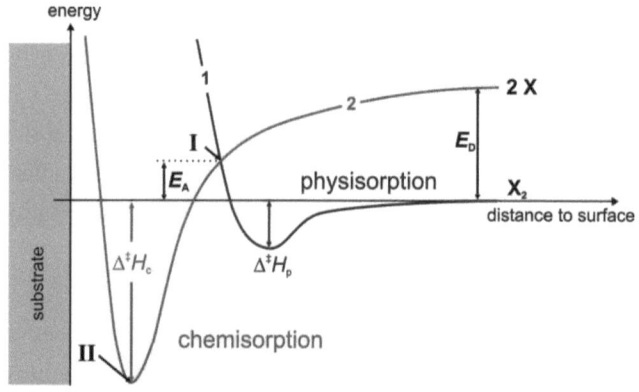

Figure 4.5: Schematic diagram of the potential energy of molecule X_2 and 2 X, respectively, as a function of distance from the surface. The flatter curve represents the molecular physisorption where the enthalpy is negative. After the point of intersection I, where an activation energy is required, the dissociative chemisorption starts. Point II represents the optimal bonding distance of X and substrate. E_D represents the dissociation energy, and E_A the activation energy for adsorption. According to the position of the point of intersection, chemisorption can require an activation $E_A > 0$, adapted from [211].

For the adsorption of H_2 on a Ni surface, the molecular hydrogen is physisorbed at an equilibrium distance of 3.2 Å while the corresponding enthalpy ΔH_P is $-4\,\text{kJ}\,\text{mol}^{-1}$. After passing the point of intersection, highlighted with I, the dissociative chemisorption starts and Ni-H bonds are formed. The chemisorption process reaches its minimum energy (point II, $\Delta^\ddagger H_C = -46\,\text{kJ}\,\text{mol}^{-1}$) at a distance of 1.6 Å. The total number of bonds is increased during this process.

¶In comparison, the distance between adjacent "physisorbed" planes of graphite is 3.35 Å.

Reactions on surfaces – Catalytic properties In the following, the process of chemisorption is exclusively restricted to metal surfaces. The Newns-Anderson model is used to elucidate the bonding situation between the adsorbate and the Fermi sea in the substrate.[212–214] Herein, the adsorbate states interact with the continuum of states in the valence band of the metal. In the case of an interaction of the adsorbate with a broad metal s or p band, the local density of states of the adsorbate is only broadened into a Lorentzian shaped state often referred to as "weak chemisorption" (see Figure 4.6 (a)). In contrast, an interaction of the adsorbate states with localized d states causes the formation of sharp bonding and anti-bonding adsorption states above and below the initial levels referred to as "strong chemisorption" comparable to chemical bonds between atoms (see Figure 4.6 (b)). Accordingly, the coupling of the d states can be described as a two-level problem.

The focus within the framework of this thesis lies on the adsorption of organic molecules on transition metals, which have a broad half-filled s band and a very narrow d band.[215] Depending on the transition metal, the d states vary in their degree of filling. Hence, the d bands can be characterized by the position of their band center ε with respect to the Fermi level, as illustrated in Figure 4.6 (c). Since the contribution of the s states to the coupling is approximately the same for each of the considered transition metals, the d states are the determining factor.[215] As moving to the left of the periodic table of elements starting with the group 11 elements, the d bands move up in energy. Thus, the anti-bonding adsorbate-metal d states become increasingly emptier, which raises the bond strength of the newly formed interaction. This renders Cu, Au, and Ag less reactive than early transition metals, where more antibonding adsorbate-metal d states are above the Fermi level. The further down in a group of the periodic table of elements the weaker the interactions with the (electronegative) adsorbates are. This downward drift in the adsorption strength from copper to gold is mainly caused by Pauli repulsion which originates from an overlapping of adsorbate orbitals with d states. Taking into account the degree of filling of the anti-bonding states of adsorption and the degree of orbital overlap with the adsorbate, gold is the most noble metal.[216]

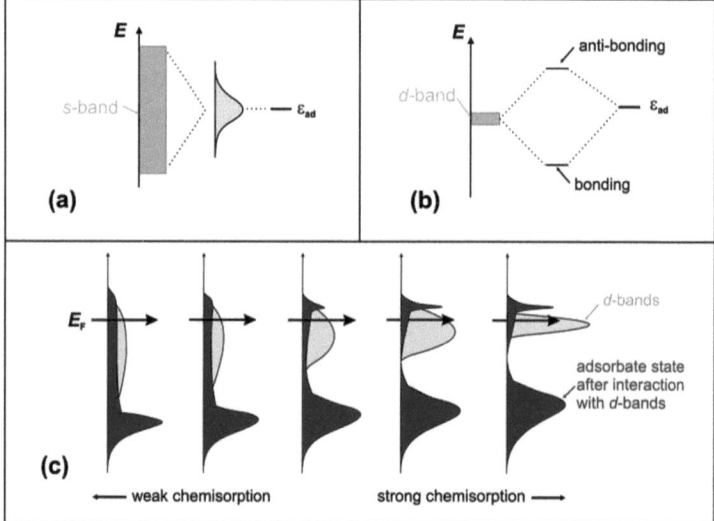

Figure 4.6: (a) and (b) The schemes illustrate the local density of states at an adsorbate for two different situations. The left sketch (a) shows the interaction of an adsorbate with a broad surface s band. In the right graphic (b) the interaction of the adsorbate with an narrow metal d band is depicted, after [215]. (c) Sketch of the local density of states projected onto an adsorbate state interacting with the d bands at a surface. Herein, the interaction strengths of the adsorbate with the surface is kept fixed as the center of the d bands ϵ_d is shifted up towards the Fermi energy ($E_F = 0$). In order to keep the number of electrons in the bands constant the width W of the d bands is decreased. Therefore ϵ_d shifts up, leading to empty anti-bonding states above E_F which cause stronger bonds. [215]

Thus, the used materials and also their surface orientation play a pivotal role for chemisorption process. Moreover, the choice of material frames the type and strength of interaction between surface and adsorbate and can either support or hinder catalytic reactions. In addition, the number of available active sites (size of the terraces, the number of edges, kinks, and available adatoms) can affect the catalytic properties of the surface.[217] In the case of not uniformly distributed active sites, the reaction can be restricted to local areas, such as kinks and step edges. So, literature supports that chemical activity is enhanced by step edges as compared to terraces.[217–219]

CHAPTER 5

Two-dimensional molecular structures

This chapter gives a summary of the publications underlying this thesis on 2D molecular networks. It is complemented by yet unpublished results. In section 5.1 the focus is placed on self-assembled 2D nanoporous networks, particularly on the dynamics of host-guest systems at the liquid-solid interface. Section 5.2 discusses the stability of such 2D networks with respect to the bonding type involved, *i.e.* halogen bonds or hydrogen bonds. The emphasis of section 5.3 lies on organic molecules which are self-assembled on graphene terminated SiO_2 or Cu foil substrates. Section 5.4 deals with a further class of self-assembled structures, namely metal-organic frameworks (MOFs). New insights of thiolate-copper coordinated networks are presented. The final section 5.5 presents the results on covalent organic frameworks (COFs).

5.1 Supramolecular host-guest networks at the liquid-solid interface*

Nanoporous 2D networks have attracted substantial interest, because of their ability to incorporate guests in exisiting pores of these networks.[92] The two dimensional host-guest networks comprise the idea of molecular recognition by

*published[92]

non-covalent interactions. In 1894, Hermann Emil Fischer described a fundamental principle for host-guest chemistry, the key-lock principle. Herein, the host and guest molecules must have mutual spatial and electronic complementarity in order to form unique complexes. Thus, the characteristic of the host matrix provides a high selectivity for guest molecules.

Owing to the tendency of all systems to minimize the overall Gibbs free energy (see section 4.1), the additional interactions between host and guest structures lower the Gibbs free energy compared to the sum of energies of the separated individual host and guest molecules. Thus, host structures are thermodynamically stable although it is opposed to nature's tendency for dense packing. The host structures are typically sustained through either hydrogen-bonds,[188] metal-ligand,[8] alkyl chain interdigitation,[220] or even van der Waals interaction.[221] The related cavity sizes range from 1 nm to 5 nm for single- or multicomponent systems.[160] Recently, also covalent porous structures were reported, which also bear the potential to act as a host matrix.[222,223]

In recent studies on host-guest systems, the main focus has been either on structure determination of the empty host matrix or the finally occupied pores.[160,224,225] Consequently, little is known about the incorporation dynamics of molecular guests into 2D supramolecular host-networks. To shed light on these processes, a standard STM setup was extended by an injection add-on (see section 2.3) which allows image acquisition while adding an additional solution containing the guest molecules. At the beginning, scanning tunneling microscopy was used to characterize the initial structure of the host matrix with submolecular resolution. These supramolecular host networks were prepared from saturated solutions of 1,3,5-benzenetricarboxylic acid (trimesic acid, TMA) or 1,3,5-benzenetribenzoic acid (BTB, synthesized following a known literature procedure[226]) in either heptanoic acid (7A) or nonanoic acid (9A). As a suitable guest, coronene (COR), *i.e.* a planar polycyclic aromatic hydrocarbon, was applied to the pre-existing film. Utilizing UV-Vis spectroscopy, the solubility of COR was determined to $(1.0 \pm 0.2)\,\mathrm{mmol\,L^{-1}}$.

After the self-assembly process of the TMA or BTB molecules on the graphite surface has taken place, topologically similar porous honeycomb structures were

Supramolecular Host-Guest Networks – Incorporation Dynamics

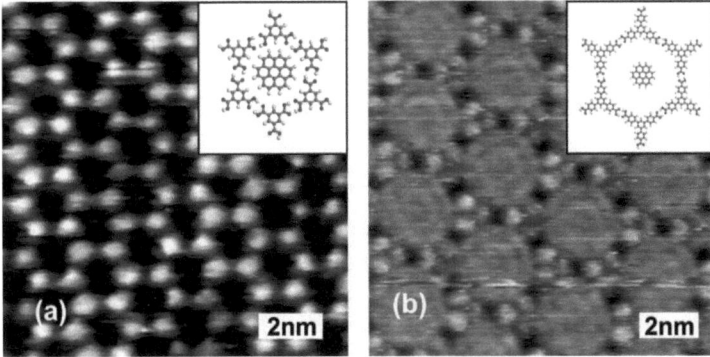

Figure 5.1: STM topographs of unoccupied host networks resulting from the self-assembly of (a) trimesic acid (TMA) and (b) 1,3,5-benzenetribenzoic acid (BTB). STM tunneling conditions: $V_T = 0.86$ V, $I_T = 57$ pA, and $V_T = 0.48$ V, $I_T = 48$ pA, respectively. The inset at the upper right corner of the images represents a structural model of the corresponding host-guest structures and exemplifies the size relation of the six-membered pore in relation to the COR guest molecule.[92]

verified with submolecular resolution by means of STM (see Figure 5.1). These networks, in which a cavity is comprised of six planar adsorbed host molecules, are stabilized by cyclic double hydrogen bonds engaged in a complementary manner. Due to the different sizes of the organic backbone of TMA and BTB, the lattice constant (1.7 nm for TMA *vs.* 3.2 nm for BTB) and accordingly the pore size, *i.e.* 1.0 nm for TMA *vs.* 2.8 nm for BTB, differs. To study the dynamics of the incorporation process, 2.5 µL of solution containing the COR molecules was applied onto the existing host-matrix.

The experimental results for TMA dissolved in 9A are depicted in Figure 5.2. The first part of the STM topograph image clearly shows the self-assembled TMA host matrix, where the pores are unoccupied. A horizontal green arrow marks the position of the application of additional solution containing the COR guest molecules dissolved in 9A as well.[†] The slow-scan direction is denoted by a vertical arrow. For the TMA pores, the geometrically matched COR molecules appear as fuzzy and streaky features, which were identified to be a transient intermediate adsorption state of the guest molecules (see Figure 5.2 (b)).

[†]90 COR molecules are offered per cavity

Two Dimensional Molecular Structures

Albeit the COR molecules are centered above the TMA host pores, the guest molecules are not directly incorporated, but rather appear considerably wider than anticipated from their geometric dimensions. However, after a time span of a few minutes, the already known contrast for the binary COR and TMA network was achieved, verifying an entire incorporation of the COR molecules (see Figure 5.2 (c)). Submolecular details of the finally adsorbed COR molecules suggest an immobilization of COR within the pores.

Nevertheless, the dimensions of the TMA host-network were unimpaired by the incorporation of the guest, *i.e.* lattice constant, orientation, and lateral extension of the domains were not affected. Varying the concentration of the guest molecules leads to a distinct difference in the incorporation dynamics. For low concentration, not all pores were occupied. Moreover, the adsorption sites follow a non-random distribution, where the COR molecules tend to cluster. In addition, the arrival rate of the guests is dependent on the concentration, because the COR molecules have to diffuse through the liquid film before getting adsorbed in the pores.

In contrast, the incorporation into the pores of the larger BTB networks proceeds entirely different. Here, several COR molecules were adsorbed within 0.5 s, whereas the intra-pore contrast did not change over time (see Figure 5.3 (a)). A comparison of the apparent heights between initially empty and occupied pores verifies the adsorption of COR molecules. Interestingly, submolecular resolution was exclusively achieved for the BTB host matrix but never for any occupied pore. This can be attributed to the high mobility of the guests rather than to physical limitations of the imaging process. According to molecular dynamics (MD) simulations, the incorporation of three guest molecules into one cavity is more stable than the adsorption of the maximum number of four guest. This holds true for both, entropy and enthalpy arguments. These MD findings also underpin the lack of resolution within the pores, due to missing lateral immobilization of the COR molecules.

Furthermore, molecular mechanics (MM) was used to derive the binding energy between COR molecules and the six-membered TMA host-matrix that is responsible for the immobilization of the guests. MM reveals that an incorporated

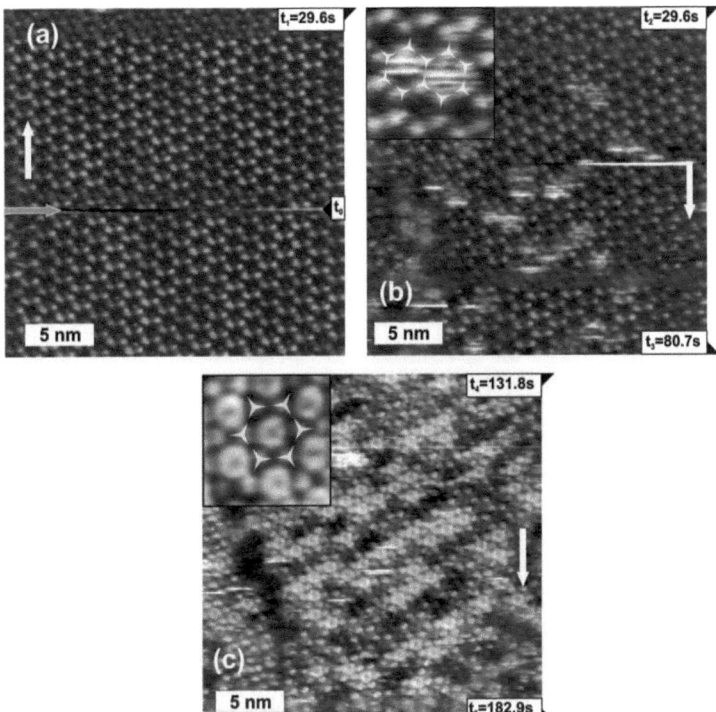

Figure 5.2: STM topographs of the incorporation process of COR molecules in the TMA host networks in 9A. (a) Initial STM topograph of the prefabricated TMA host network, where the position of the guest injection is marked by a green horizontal arrow. (b) STM topograph showing the transient intermediate adsorption state of the COR molecules. (c) STM topograph with submolecular resolution illustrates the final adsorption state, where guest molecules were incorporated into the TMA host-matrix. The inset in the STM topographs presents detailed views of COR guests in the intermediate state and in the final adsorption state, respectively. In each case, the TMA host network is indicated by light-gray triangles. STM tunneling conditions: $V_T = 0.73\,\text{V}$, $I_T = 71\,\text{pA}$, and $T = 297.0\,\text{K}$ for all topographs.[92]

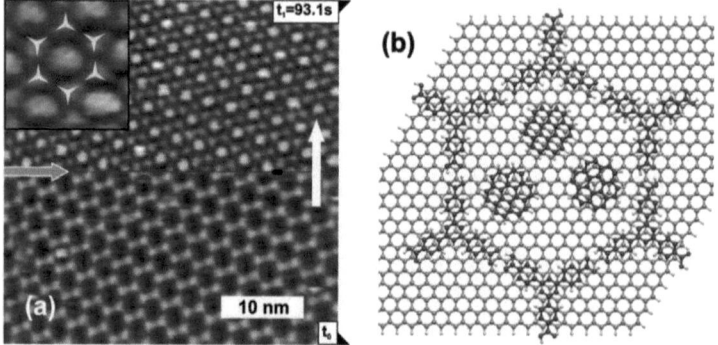

Figure 5.3: STM topograph of COR incorporation into BTB host networks in 9A ($V_T = 0.76$ V, $I_T = 51$ pA, $T = 297.0$ K). The slow-scan direction is marked by the vertical white arrow and the horizontal green arrow indicates the scan line where saturated COR solution was added; a slight disturbance is also visible. The intrapore contrast changes within one to three scan lines (*i.e.* in less than 0.5 s). The inset shows a detailed view of an occupied pore, and the six bordering BTB molecules are indicated by light-gray triangles. (b) Geometry-optimized model of three COR molecules incorporated within one BTB cavity. Although up to four COR molecules would fit into one BTB cavity, MD simulations indicate that this situation is unstable and one COR desorbs. However, because of the lateral mobility of the loosely packed three remaining COR molecules within the BTB pore, molecular resolution cannot be obtained. Thus, it is not possible to verify the postulated structure experimentally.[92]

COR molecule is stabilized by 18 hydrogen bonds with the pore wall (6 C-H···$O_{carbonyl}$ and 12 C-H···$O_{hydroxyl}$ hydrogen bonds, see Figure 5.4 (a)). In order to determine the bonding strength of the additional hydrogen bonds, the distance dependent adsorption energies[‡] of a COR molecule accommodated in a TMA cavity and just above pristine graphite were calculated. Figure 5.4 presents the adsorption energies of COR adsorption above pristine graphite (red curve) and in a six-fold TMA pore on graphite (blue curve) derived from MM. Each of these hydrogen bonds contributes on average $2.2\,kJ\,mol^{-1}$, while the bonding distance of the C-H···O hydrogen bonds lies approximately at 2.74 Å. However, the normal bonding lengths for C-H···O hydrogen bonds amount to 2.5 Å causing higher binding energies in the range of $4\,kJ\,mol^{-1}$ to $8\,kJ\,mol^{-1}$.[223]

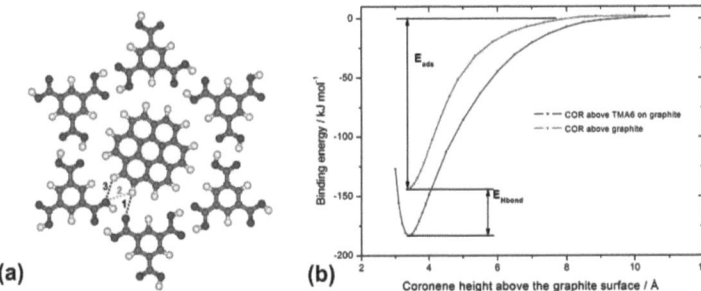

Figure 5.4: (a) Sketch of the final structure of a TMA COR host-guest network that is stabilized by 18 additional hydrogen bonds, *i.e.* 6 C-H···$O_{carbonyl}$ and 12 C-H···$O_{hydroxyl}$, where the hydroxyl oxygen atom accepts two hydrogen bonds. Three hydrogen bonds are indicated. (b) adsorption energies of COR above pristine graphite (red curve) and within a six-fold TMA cavity again above graphite (blue curve). The additional stabilization of the COR molecules leads to a distance-dependent energy difference.[92]

Besides the influence of the geometry, the used solvent also affects the incorporation dynamics. For the TMA network dissolved in heptanoic acid, the transient intermediate state was observed very rarely in contrast to TMA networks dissolved in nonanoic acid. It is generally accepted that the pores at the liquid-solid interface are not empty but are occupied by solvent molecules. Thus, the solvent molecules within the pores need to be displaced for the guest incorporation. The dynamics of this process is much faster for heptanoic acid

---

[‡]Binding energies of geometry optimized structures, derived by molecular mechanics calculations.

than for nonanoic acid and for larger BTB pores than for smaller TMA pores. Details can be found in the attached publication.

5.2 Halogen versus hydrogen bonded 2D networks

In contrast to hydrogen bonded 2D networks, which have been studied extensively, just a few studies are currently available on halogen stabilized self-assembled networks.[192,227,228] The major reason for that is the minor distribution of halogenated compounds, which are the basis for halogen bonds. Yet, a combination of halogen and hydrogen bonds, that are simultaneously stabilizing a network, is possible and not unlikely, as recently reported by Yoon *et al.*[229] This is not surprising because both types of interactions are based on a similar bonding mechanism, *i.e.* a donor and acceptor principle as illustrated in section 4.2.

The molecules of choice were composed of the same organic backbone, *i.e.* four phenyl rings, but functionalized either with the already studied carboxylic acid group -COOH or iodine atoms I. The 1,3,5-benzenetribenzoic acid (BTB) offers potential binding sites for hydrogen bonds, whereas the 1,3,5-tris(4-iodophenyl)benzene (TIPB) bears the potential for halogen stabilized networks. In order to characterize and compare the strength of these two fundamental types of intermolecular interactions, STM studies were conducted on the graphite-fatty acid interface. HOPG, a non-catalytic and unreactive surface for these type of molecule, was chosen to minimize molecule-substrate interaction. Thus, the final structure is governed by molecule-molecule interactions rather than by molecule-substrate interactions.

In our first series of experiments, the self-assembly of saturated TIPB solution deposited onto a freshly cleaved graphite surface was investigated. The finally observed porous 2D network structure was stable against changes of the chain length of the fatty acid. Figure 5.5 (a) - (c) present STM topographs of routinely obtained TIPB structures dissolved in heptanoic acid (7A), octanoic acid (8A), and nonanoic acid (9A), respectively.

It is worthwhile mentioning that deposition of TIPB onto more reactive surfaces such as coinage metals enables the scission of the carbon halogen bond and initiate a covalently interlinked structure.[230,231] In the following, the solvent is restricted to nonanoic acid to assure comparability and avoid introducing an additional parameter. A detailed analysis of the STM topograph shown in

TWO DIMENSIONAL MOLECULAR STRUCTURES

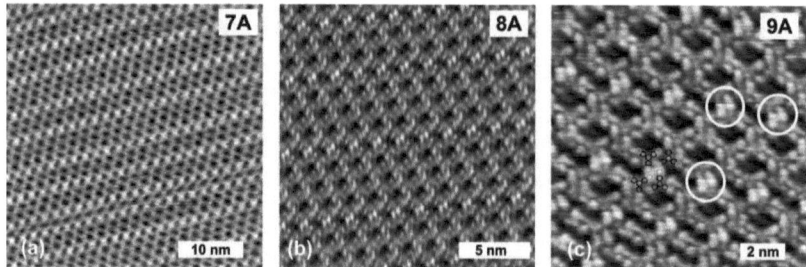

Figure 5.5: STM topographs of 1,3,5-tris(4-iodophenyl)benzene (TIPB) on HOPG in heptanoic acid (a), octanoic acid (b), and nonanoic acid (c), respectively. The bright features in the STM tunneling contrast are caused by a high concentration of iodine atoms. Exemplarily, four iodine atoms in image (c) are encircled white. Independent of the used solvent, similar features are obtained in all STM topographs, suggesting a similar basic binding motif. STM tunneling conditions: (a) $V_T = 0.59\,\text{V}$, $I_T = 55\,\text{pA}$; (b) $V_T = 0.87\,\text{V}$, $I_T = 65\,\text{pA}$, and (c) $V_T = 0.74\,\text{V}$, $I_T = 46\,\text{pA}$, respectively.

Figure 5.5 (c) reveals that the TIPB structure is stabilized by a complicated mixture of halogen-halogen interactions (encircled white) as well as halogen-hydrogen interactions. An exclusive stabilization *via* halogen bonds, as reported for the analog TBPB molecules under ultra-high vacuum conditions at 80 K, can be excluded.[232] The geometrical extent of the molecules indicates an intact adsorption on graphite which is consistent with literature.[233] When applying a positive tunneling voltage to the tip, the iodine atoms terminating the threefold TIPB molecules appear as bright protrusions. The obtained contrast can be explained in terms of tunneling from filled states of the electron-rich iodine atoms and phenyl rings (HOMO) to unoccupied states of the tip, as it was explained for halogen substituted phenyl octadecyl ethers.[234] The variation of the distances between the bright rows, *i.e.* high iodine concentration, points to missing long range ordering. Yet, the occurrence of these remarkable features in every STM topograph, independent of the used solvent, suggests similar binding motifs, even submolecular resolution was exclusively obtained in nonanoic acid.

The BTB networks have been studied intensively as presented in section 5.1 and elsewhere.[101,235] It is common knowledge that each of the carboxylic acid groups of a BTB monomer forms a double hydrogen bond with the adjacent monomer, when using nonanoic acid as solvent. According to the three-fold symmetry of the building block, hexagonal porous networks with a lattice parameter of 3.2 nm were formed.

Besides the presented studies of pure TIPB and BTB networks, also a mixture of both molecules was investigated. Therefore, a certain volume ratio of saturated solutions was applied onto a graphite surface. By varying the ratio of these different molecules, the competitive assembly behavior between halogen-bonded and hydrogen-bonded networks was studied. In the case of an excess of BTB, *i.e.* a volume ratio of 1:4 for the benefit of BTB, the known self-assembled BTB honeycomb structure was obtained as depicted in Figure 5.6 (a).

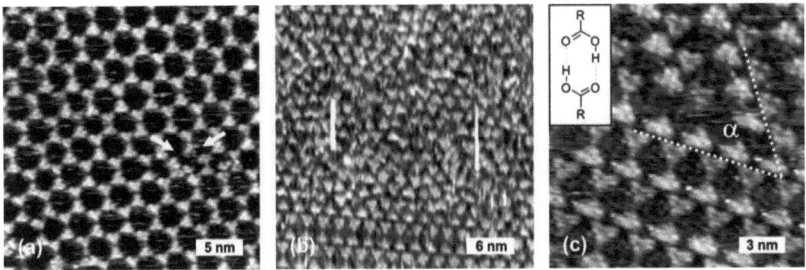

Figure 5.6: STM topographs of a mixture of TIPB and BTB with different ratios, *i.e.* (a) 1:4, (b) 3:1, and (c) 10:1. (a) The honeycomb structure is disturbed by some defects, highlighted by white arrows. (b) This STM topograph shows a mixture of BTB row structure and some disordered areas. (c) Perfect BTB row structure with two different orientations, indicated by white dashed lines, $\alpha = (60 \pm 2)°$. The inset presents the double hydrogen bond, the basic binding motif for each structure. STM tunneling conditions: (a) $V_T = 0.70\,\text{V}$, $I_T = 0.55\,\text{pA}$; (b) $V_T = -0.82\,\text{V}$, $I_T = 0.55\,\text{pA}$, and (c) $V_T = 0.69\,\text{V}$, $I_T = 63\,\text{pA}$.

In contrast to other studies exclusively using BTB molecules, the mixture of TIPB and BTB tends to result in a higher defect density, as highlighted by white arrows in Figure 5.6 (a). Through the intrinsic self-repair mechanism of these networks, the misaligned BTB molecules were reoriented and thus the defects were recovered. Increasing the ratio up to 3:1 for the benefit of TIPB, a mixture between the BTB row structure and some disordered areas was obtained (Figure 5.6 (b)). At this point, any contribution of TIPB to the final structure can be excluded due to missing bright contrast features, which would point to iodine atoms. Moreover, no lateral offset between two TIPB molecules that are forming the basic dimer could be revealed. Increasing the ratio to 10:1 for the benefit of TIPB leads exclusively to a BTB row structure as shown in Figure 5.6 (c). The main axes of the highlighted rows span an angle of $\alpha = (60 \pm 2)°$. Both the honeycomb as well as the row structure are composed of the same basic BTB dimer motif, where two BTB monomers are stabilized by a double hydrogen bond as sketched in the inset of Figure 5.6 (c). Unfortunately, the accurate interaction scheme between two BTB dimers within the row structure cannot be revealed. This was already reported by Kampschulte et al.[101] Within the investigated range of TIPB:BTB volume ratios, the hydrogen bonded structure was favored over the halogen stabilized structures. Surprisingly, an increase of the ratio for the benefit of TIPB triggers a phase transition of the BTB honeycomb structure into the more densely packed row structure rather than forming any TIPB networks. Although only weak interactions between the molecules and the graphite surface are prevailing, slightly different adsorption energies between TIPB and BTB are possible due to the different dimensions of the functional groups. To gain further details, DFT calculation of the entire system would be desirable.

5.3 Organic molecules on graphene terminated substrates[§]

Since the discovery of graphene, surface scientists have wanted to utilize its outstanding properties (see section 3.3) for applications.[236] A step forward is the use of graphene terminated surfaces as a template for the assembly of organic molecules. In 2009 Sykes reported on the formation of robust organic monolayers of 3,4,9,10-perylene tetracarboxylic dianhydride (PTCDA) on graphene terminated SiC(0001) under ultra-high vacuum conditions.[119] In the following, an STM study at the liquid-solid interface on a graphene terminated conducting Cu foil as well as on insulating SiO_2 is presented. In general, such graphene terminated surfaces are suited for all kinds of molecules which have been successfully applied on conducting HOPG surfaces so far. Submolecular resolution of aromatic molecules applied on such graphene terminated surfaces was regularly obtained.

In a first study, the self-assembly process of p-terphenyl-3,5,3',5'-tetracarboxylic acid (TPTC) has been investigated by means of STM. According to Blunt et al.[125,237] these molecules are known to form self-assembled networks on a graphite surface, where the motifs are composed of three, four, five, or six building blocks. Independent of the number of molecules involved, the bonding motifs are stabilized by directional double hydrogen bonds, resulting in two possible placements of adjacent TPTC molecules, i.e. parallel and arrow-head configuration.[125] Based on these fundamental binding motifs, the self-assembled 2D molecular networks can be described by a random, entropically stabilized rhombus tiling. For this purpose, each molecule is represented by a rhombus, as demonstrated in Figure 5.7.

In a previous study,[125] the terphenyl backbone of TPTC appeared as bright rodlike feature, but single phenyl rings were hardly resolved. However, for graphene terminated SiO_2 every binding motif predicted above was revealed with submolecular resolution. The STM topograph image shown in Figure 5.7 (a) presents an overview, where every single phenyl ring of the TPTC is resolved.

[§]Paper in preparation, Georg Eder, Izabela Cebula, and Peter Beton, University of Nottingham

Two dimensional molecular structures

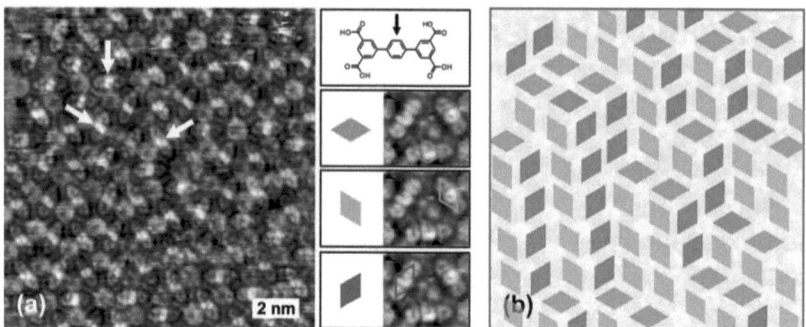

Figure 5.7: (a) STM topograph of p-terphenyl-3,5,3',5'-tetracarboxylic acid (TPTC) dissolved in nonanoic acid on graphene terminated SiO_2. Single phenyl rings are resolved. The phenyl ring in the middle clearly appears brighter. The insets represent the chemical structure of TPTC and a color-coded sketch of each of the three different orientations of the molecules. (b) Analysis of the different orientations of the STM topograph in (a). STM tunneling conditions: $I_T = 91.9\,\text{pA}$, $V_T = 1.02\,\text{V}$.

Interestingly, the phenyl-ring in the middle of a TPTC molecule appears brighter in STM tunneling contrast, independent of the binding motifs. A detailed analysis of the orientation of the molecules is presented in Figure 5.7 (b). As obvious from the STM topograph, there are three different orientations discernible, indicated by color-coded rhombi for their directions. The high resolution of these aromatic molecules on graphene terminated surfaces was assigned to the largely unperturbed electronic properties of the monolayer. This is in agreement with recent UHV studies of organic molecules on graphene.[119] These findings underpin the assumption of weak interaction between TPTC and graphene, which is just coupled by their conjugated π electron systems and omnipresent van der Waals interactions. Moreover, it is impressive that every experimental run resulted in an identical final structure on both substrates, *i.e.* insulating SiO_2 or conducting Cu foil.

Surprisingly, the entire surface was covered with TPTC molecules, indicating a defect-free intact graphene layer. To verify a single graphene layer on the substrate, Raman spectroscopy was used, as demonstrated in section 3.3. In addition, the detection of a tunneling current within the graphene layer on top of insulating SiO_2 supports at least an electrically conductive graphene layer.

Nevertheless, the graphene sheets are not flat and possess a lot of wrinkles as obvious from the height modulation in Figure 5.7 (a). The STM topograph images of TPTC on graphene terminated Cu foil also confirm uninterrupted graphene (see Figure 5.8 (a)). Again, submolecular resolution could be achieved. As reported in literature, these graphene sheets are flowing over the step edges and thereby connecting adjacent terraces at the cost of not perfectly flat surfaces.[82]

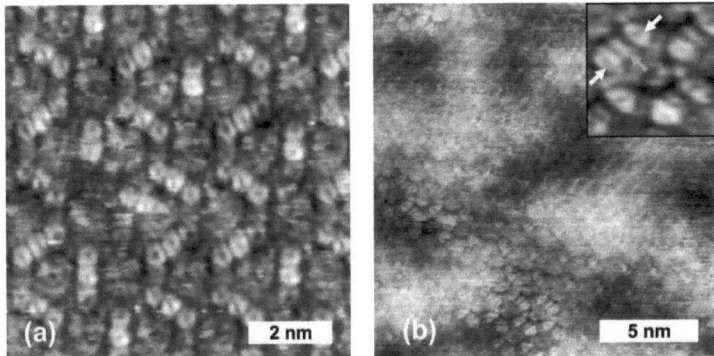

Figure 5.8: (a) STM topograph of p-terphenyl-3,5,3',5'-tetracarboxylic acid (TPTC) dissolved in nonanoic acid on graphene terminated Cu foil. (b) STM topograph of 1,3,5-Benzenetricarboxylic acid (TMA) dissolved in nonanoic acid on graphene terminated SiO_2. The inset clearly shows a hexagonal pore, which is composed of six TMA molecules. The phenyl-rings are resolved as donut like structures and highlighted by white arrows, whereas a hydrogen bond is exemplarily marked by a blue arrow. STM tunneling conditions: (a) $I_T = 50\,\text{pA}$, $V_T = 0.60\,\text{V}$ and (b) $I_T = 102\,\text{pA}$, $V_T = 0.76\,\text{V}$.

Furthermore, the experiments on graphene terminated SiO_2 were extended to a standard system in STM, *i.e.* TMA dissolved in nonanoic acid, as depicted in the STM topograph of Figure 5.8 (b). Here, only in some regions submolecular resolution could be obtained owing to the poor graphene quality, *i.e.* a wavelength of the graphene wrinkles in the 10 nm regime. Nevertheless, the inset of Figure 5.8 (b) clearly shows a six membered pore of TMA molecules. The donut-like features forming these hexagonal pores can be addressed to the single phenyl-rings of the TMA molecules. In between two adjacent molecules some contrast features, representing double hydrogen bonds were observed.

5.4 2D metal-organic frameworks (MOFs) based on thiolate-copper coordination bonds[¶]

The interaction strength within metal-organic frameworks occupies an intermediate position between hydrogen-bonded and covalent organic frameworks,[8,28,239] and thus combines the advantages of both, *i.e.* overall stability and bond reversibility. The reversibility of the bonds allows for the growth of long-range ordered structures, also extended in the third dimension. Metal-organic frameworks are candidates for molecular recognition catalysis, gas storage materials, and separation techniques. To date, a great variety of structural and chemical properties of surface-supported metal-organic networks has been reported.[28,239,240]

Recent studies have often been focused on surface-confined coordination networks based on copper-carboxylate coordination bonds.[201,241] In the following, the toolbox of surface-supported metal-organic networks is extended by thiolate-metal complexes under ultra-high vacuum conditions. After a Cu(111) single-crystal was prepared by cycles of Ar$^+$-ion sputtering and subsequent annealing at 820 °C, the organic monomer 1,3,5-tris(4-mercaptophenyl)benzene (TMB) was thermally evaporated from a home-built Knudsen cell with crucible temperatures of approximately 145 °C. To characterize the samples, scanning tunneling microscopy was applied. Figure 5.9 (a) shows an STM topography image of the self-assembled and densely packed trigonal structure of TMB on a Cu(111) surface, as obtained by room temperature experiments.

A detailed analysis of the lattice parameter of the structure, *i.e.* (1.30 ± 0.05) nm, verifies the deprotonation of the thiolated building block. Moreover, it is generally accepted that the adsorption of thiols on reactive metals leads to deprotonation and subsequently the generated thiolates are anchored by newly formed sulfur-metal bonds.[242–244] The three sulfur-copper bonds stabilize a planar adsorption geometry, while the commensurate adsorption position of the TMB on the substrate suggests that all sulfur atoms are very likely to reside at similar adsorption sites.

[¶]published[238]

2D METAL-ORGANIC FRAMEWORKS (MOFS)

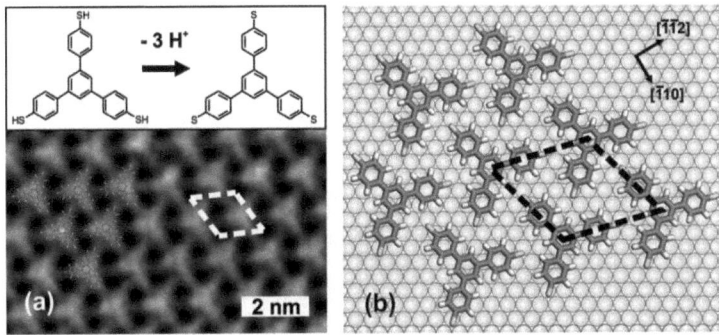

Figure 5.9: (a) STM topograph of TMB on Cu(111) acquired at room temperature ($V_T = 0.79$ V, $I_T = 185$ pA, $a = b = 1.30$ nm, $\gamma = 120°$, unit cell indicated by dashed white lines), where a densely packed trigonal structure is revealed. The upper part of the image sketches the deprotonation of TMB when adsorbed on reactive copper surfaces and forming a surface-anchored trithiolate. (b) Tentative model of the densely packed trithiolate structure including the Cu(111) surface. While the azimuthal orientation of the TMB-derived trithiolates with respect to the substrate directions can be inferred from the experiment, the precise adsorption site is not known.[238]

Subsequently, the self-assembled TMB structures were thermally annealed at 160 °C to 200 °C for approximately 10 min. This converts the trithiolate monolayer into two polymorphs of metal-organic coordination networks. Here, the required metal coordination centers were supplied by the free adatom gas of the Cu(111) surface.[245,246] Both, the hexagonal honeycomb structure and a centered rectangular dimer row structure are depicted in Figure 5.10. This conversion is accompanied by reorientation and repositioning of the TMB molecules and the introduction of copper coordination centers. Heating of the sample to higher temperatures leads to an increase of the area density of the free copper adatom gas, which is a temperature dependent evaporation/condensation equilibrium of Cu atoms at step edges. The additionally applied thermal energy also enhances the lateral mobility of the species and enables the formation of new molecular networks.

An evaluation of STM topographs of the hexagonal structure (see Figure 5.10 (a)-(c)) reveals a lattice constant of 3.4 nm and the plane space group *p6mm*. A model, that coordinates each functional group in a linear manner *via* two

Two dimensional molecular structures

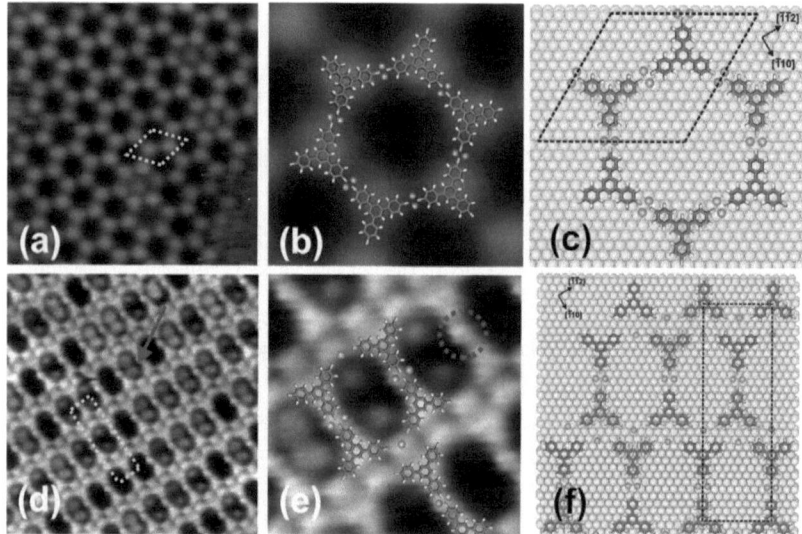

Figure 5.10: (a) STM topograph ($24 \times 24\,\text{nm}^2$, $V_T = 1.0\,\text{V}$, $I_T = 67\,\text{pA}$) and (b) close-up ($6.5 \times 6.5\,\text{nm}^2$) of the honeycomb structure. The unit cell is indicated by dashed lines, i.e. $a = b = 3.4\,\text{nm}, \gamma = 120°$. (c) Tentative model of the TMB honeycomb structure on a Cu(111) surface. (d) STM topograph ($18 \times 18\,\text{nm}^2$, $V_T = 0.8\,\text{V}$, $I_T = 121\,\text{pA}$) and (e) close-up ($6.9 \times 6.9\,\text{nm}^2$) of the less symmetric dimer row structure. The unit cell is indicated by dashed lines, i.e. $a = 2.2\,\text{nm}, b = 6.6\,\text{nm}, \gamma = 120°$. The bright features between the TMB molecules (highlighted by a red dashed circle) are attributed to Cu adatoms, indicating a metal coordinated network. (f) Tentative model of the dimer row structure on top of the Cu(111) surface.[238]

2D Metal-Organic Frameworks (MOFs)

Cu adatoms, meets these requirements. In contrast, the less symmetric row structure is composed of rows of dumbbell-shaped dimers, as illustrated in Figure 5.10 (d)-(f). In this motif, a dimer itself is coordinated *via* two Cu atoms – which is the same basic structural motif as in the honeycomb structure. Yet, the dimers are coordinated just by a single Cu adatom with respect to each other. Between the dimers of adjacent rows, an offset by half a lattice parameter along the row axis was observed, as demonstrated in Figure 5.10 (e). As clearly evident from the STM topographs and indicated by a red arrow (see Figure 5.10 (d)), some TMB molecules were entrapped during this conversion and gave rise to additional contrast features within the pores. Nevertheless, both structures were observed in coexistence. The ratio of the phases is dependent on the initial coverage of the self-assembled precursor structures. A higher initial coverage develops the row structure, which is more densely packed in contrast to the hexagonal structure.

To gain a fundamental understanding of the thiolate-copper coordination bonds and to confirm the tentative binding models mentioned above, DFT calculations were performed. For computing cost reasons, the system was reduced to the connecting nodes, modeled by two phenylthiolates and corresponding copper centers. The influence of the substrate was neglected. Figure 5.11 presents four different simplified intermolecular bonding schemes. Here, metal-coordination bonds mediated by one or two copper atoms and a covalent disulfur bridge were simulated. Only the two-center coordination bond results in a linear connection of the molecules. One center coordination bond, independent of the fact whether it is *trans*-coordination and *syn*-conformation or *trans*-coordination and *anti*-conformation, and disulfur bridges result in lateral offsets perpendicular to the bond axis.[||]

According to DFT results, the two-center coordination complex is the strongest with a binding energy of $555\,\mathrm{kJ\,mol^{-1}}$ compared to $394\,\mathrm{kJ\,mol^{-1}}$ for *trans*-coordination and *syn*-conformation, $397\,\mathrm{kJ\,mol^{-1}}$ for *trans*-coordination and *anti*-conformation, and $151\,\mathrm{kJ\,mol^{-1}}$ for disulfur bridges. Yet, since no lateral

---
[||]Trans-coordination means "across" or "on the other side" of the main coordination plane, syn-conformation means on the same side, anti-conformation means on the opposite side.

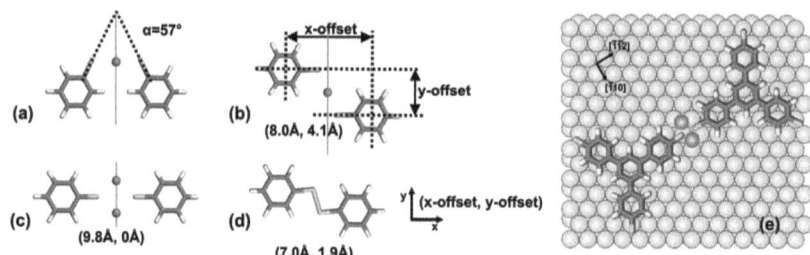

Figure 5.11: Sketches of DFT geometry optimized intermolecular bonding schemes of interconnected phenylthiolates *via* (a) one-center trans-coordination syn-conformation ($\alpha = 57°$), (b) one-center trans-coordination anti-conformation (x-offset = 8.0 Å, y-offset = 4.1 Å), (c) two-center coordination bond (x-offset = 9.8 Å, no y-offset), (d) covalent coupling *via* disulfur-bridge (x-offset = 7.0 Å, y-offset = 1.9 Å). The numbers in brackets give the center-to-center distance of the phenyl groups (x-offset) and the perpendicular axial offset (y-offset), respectively. (e) Tentative model of the fundamental building unit, *i.e.* two-copper coordinated TMB dimer, adsorbed on the Cu(111) substrate. The sketch shows similar adsorption lattice sites for both coordinating copper atoms (colored red) to simultaneously optimize their adsorption energy.[238]

displacement along the dimer axis was detectable in all STM topographs, DFT calculations suggest the coordination *via* two copper atoms. Moreover, the experimentally obtained bond lengths confirm this assumption. Consequently, the two-center coordination bond motif is the only basic motif which is involved in the formation of the honeycomb structure. In order to explain the row structure, a more complex model involving different types of intermolecular bonds is proposed. Besides the dimer motif, which is involved in both structures, STM topographs reveal contrast features directly above and below intra-row neighbors, suggesting a *trans-syn*-arrangement. Figure 5.11 (e) depicts the adsorption geometry of a dimer on the Cu(111) surface. The axes of the dimers are oriented along the $\langle \bar{1}\bar{1}2 \rangle$ direction of the Cu surface. This model illustrates the adsorption of the copper atoms at similar sites, for instance as tentatively shown in three-fold hollow sites.

In order to address the influence of the type of substrate material, the same preparation protocols were applied to Ag(111). For room temperature deposition of TMB a densely packed trigonal structure was obtained, if the precursor was already deprotonated similar to the Cu(111) case. Following the protocol, the sample was annealed resulting in distinctly different structures compared to the Cu(111) substrate. Annealing the sample up to 250 °C for 1 h did not result in a phase transition. However, further annealing of the Ag(111) sample up to 300 °C for 1 h results in disordered glassy networks in coexistence with remnants of the precursor structure. Some of the irregular structures can be identified as covalently interconnected monomers. The bonding distances within the interconnected monomers are in good agreement with disulfur bridge distances calculated by DFT. Here, for computing time reasons the substrate was neglected. In general, a high degree of disorder is an evidence for covalent networks, in which the irreversible covalent bonds are formed during the growth and cannot be corrected afterwards. The absence of metal coordination bonds on the Ag(111) surface can be explained by the different affinity of Cu *vs.* Ag adatoms to form metal-thiolate coordination bonds.[247]

5.5 2D Covalent Organic Frameworks – COFs

2D polymers are expected to be a next generation material in several areas of engineering. They are composed of one monomer thick, covalently bonded molecular sheets with a long-range ordered (periodic) internal structure.[248] The most common example of synthetic 2D polymers is graphene, an only one carbon atom thick, covalently bonded, long-range ordered thin film of carbon. For generating 2D networks with a single type of compound, molecules that provide three bonding sites are preferred. To build up such synthetic polymers, there are several different approaches available ranging from light-induced 2D polymerization, electrochemically induced polymerization, heat-induced polymerization, or condensation reactions.[205,206] In the following, the approaches are divided in two different groups corresponding to the environmental conditions, *i.e.* ultra-high vacuum and ambient conditions.

5.5.1 2D COF's under ultra-high vacuum conditions**

Here, the focus is on covalent coupling through radical addition reactions of appropriate monomers on metallic substrates. After the halogen substituted polyaromatic monomer, *i.e.* 1,3,5-tris(4-bromophenyl)benzene (TBPB), was sublimated onto either a Cu(111) or Ag(110) surface, subsequent annealing was performed to generate radicals and interlink the precursor molecules. During this process, the substrate confines the molecular motion and acts as catalyst for activating the tailored molecules.[124,249] The chosen TBPB molecule offers three predetermined breaking points at the C-Br bonds. This is evident, as the homolytic bond dissociation energy of the C-Br bond of 3.2 eV is substantially lower than the energy of the C-C bonds within the phenyl rings (4.8 eV) rendering the molecule suitable for dehalogenation.[250,251] Moreover, the Cu(111) substrate is known to facilitate the homolysis of halogen functionalized molecules.[252] In a preliminary study, the TBPB molecules were evaporated onto graphite, leading to an ordered self-assembled chain-like structure. Figure 5.12 (a) shows an STM topograph of such chains of intact TBPB dimers that are stabilized by Br···H-C

**published [232,233]

bonds. Here, every other row appears brighter. This contrast modulation implies different adsorption sites of the TBPB on the graphite surface. It is not surprising that due to the missing catalytic properties of graphite, annealing to 320 °C only leads to a complete desorption instead of triggering dehalogenation.

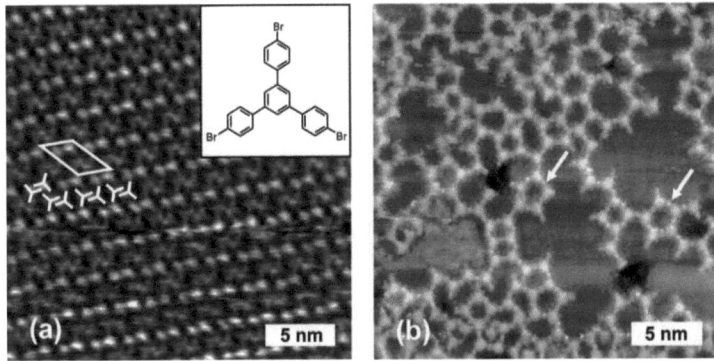

Figure 5.12: (a) STM topograph of 1,3,5-tris(4-bromophenyl)benzene (TBPB) on graphite. $V_T = 1.9$ V, $I_T = 76$ pA. The unit cell is indicated by white lines (a =3.4 nm, b =2.5 nm, α = 44°). (b) Typical STM tunneling contrast of TBPB on Cu(111) surface, when the monomers were sublimated at room temperature. The spherical protrusions, marked by white arrows, are identified as Cu adatoms that coordinate the TBPB molecules. [233]

In contrast to the results on a graphite surface, the adsorption of TBPB molecules on reactive metal surfaces kept at room temperature ends up in open-pore networks. The center-to-center distance between adjacent monomers was determined as (1.49 ± 0.10) nm for Cu(111) and (1.57 ± 0.06) nm for Ag(110). Comparing the experimentally determined center-to-center distances to typical C-C bonds, the experimentally investigated lengths are definitely larger. Figure 5.12 (b) offers an explanation by revealing bright protrusions between adjacent molecules, which were identified as single copper atoms. It is obvious from the distances that two or occasionally three TBPB molecules are coordinated *via* metal atoms even if the metal atom often remains unresolved as reported for similar systems.[249,253] On the route towards covalently interlinked networks, these protopolymers[††] are just an intermediate state. For Cu(111)

[††]Protopolymers are a new chemical state. Here, a strong interaction is formed between dehalogenated monomers, *i.e.* radicals, and metal "adatoms" on the metal surface without generating chemical bonds. [253]

Two dimensional molecular structures

and Ag(110) this homolytic bond dissociation is observed without allocating any activation energy. A promising step to convert protopolymers to 2D COFs is annealing the sample in order to release the coordinating metal atom and form covalent bonds. After the sample was annealed to 300 °C, the center-to-center distance of interconnected molecules was reduced to (1.24 ± 0.06) nm, as visible in Figure 5.13. Thus, annealing is essential for triggering the physical change within the monolayer, *i.e.* the reduction of the spacing of adjacent molecules.

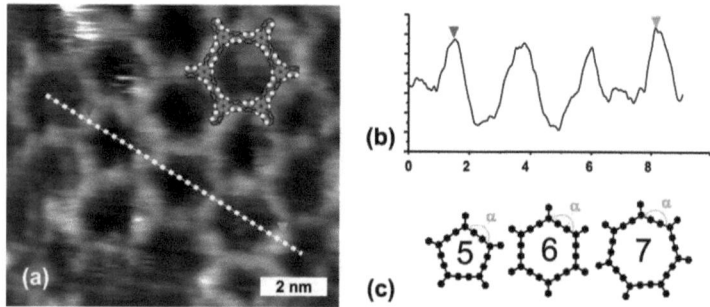

Figure 5.13: (a) STM topograph of a covalent organic framework on Cu(111), after the TBPB molecule were annealed to 300 °C. (b) Line-profile across three pores of the interlinked network. The center-to-center distance was determined to 2.2 nm. (c) Sketch of possible binding motifs: pentagon, hexagon, and heptagon. The newly formed bond angle is distorted by $\Delta\alpha$.[233]

Unfortunately, the defect density of these 2D porous networks is quite high. The optimal bonding motif of the threefold symmetric TBPB molecules would be a hexagon, where all newly formed bonds exhibit an ideal bonding angle of 180°. In the STM topographs a variety of differently membered rings were regularly observed, as sketched in Figure 5.13 (c). Here, each newly formed bond is distorted from its optimal equilibrium angle on average by $\Delta\alpha = 60°(6-N)/(3N)$, where N is the number of molecules involved in the motif. The variation from the ideal bonding angle accounts to 4°, 0°, and −2.9° for pentagon, hexagon, and heptagon, respectively. Due to small energy variations caused by non-ideal bonding angles, a high number of polygons different from the hexagons was revealed by STM.

In order to study the material- and orientation-dependent reactivity for carbon-bromine bond homolysis, further temperature-dependent STM experiments were carried out. Therefore, the substrate temperature was introduced as an additional parameter. Deposition of TBPB onto Cu(111) at 80 K leads to highly ordered, virtually defect-free self-assembled structures. Figure 5.14 (a) presents an STM topograph, where intact molecules are discernible by their three-fold symmetry. This molecular structure is based on a hexagonal lattice with $a = (2.05 \pm 0.06)$ nm where a unit cell consists of one TBPB molecule. The corresponding structural bonding motif is depicted in Figure 5.14 (b). Interestingly, the fundamental attractive interaction is based on the non-spherical charge distribution around the bromine substituents, *i.e.* a force between a positive cap opposing the C-Br bond and a negative potential around the bond axis.[197] Annealing this sample to room temperature induces the formation of protopolymers, similar to the observation for room temperature deposition. Thereby, the Cu adatoms are generated by either a continuous process of detachment from step-edges or extraction from terraces, which can occur even at room temperature.[254,255]

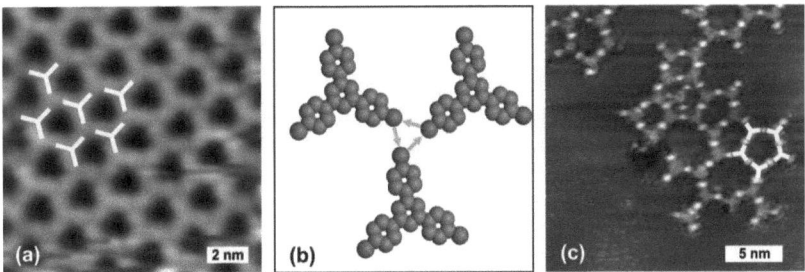

Figure 5.14: (a) STM topograph of TBPB deposited onto Cu(111) kept at 80 K, where a non-covalent self-assembled highly ordered structure is clearly visible. The threefold symmetric features are assigned to single intact TBPB molecules. (b) Tentative structure model of the TBPB molecules on Cu(111) at 80 K. The structure is stabilized by halogen-halogen interaction as a consequence of the non-spherical charge distribution of the bromine atoms. (c) STM topograph after warming the sample to room temperature.[232]

Substituting the Cu(111) surface by an Ag(110) surface leads to qualitatively similar results for room temperature deposition of TBPB, *i.e.* the formation of protopolymers. However, deposition onto the Ag(110) surface kept at 80 K does not lead to any ordered structure as observed for Cu(111). Instead, isolated and immobilized TBPB molecules were observed without any long-range ordering. For face-centered-cubic (fcc) metals, these findings can be explained by the higher surface corrugation on (110) surfaces than on (111). As a consequence of this, thermally activated surface diffusion on (110) surfaces is more easily suppressed at lower temperatures.

To gain deeper insight into the dependence on the crystallographic orientation and the material, further experiments were conducted on the Ag(111) surface. As expected for low temperature deposition on Ag(111), also a self-assembled structure similar to the Cu(111) case was observed. Surprisingly, a variety of distinct self-assembled phases based on intact molecules was present for room temperature deposition instead of protopolymers that would be a clear indication of dehalogenation. Figure 5.15 (a)-(d) show some representative structures of the TBPB on Ag(111) at room temperature, where in (a) three different coexisting phases, *i.e.* a row structure in the upper center part, a hexagonal flower structure in the upper right part, and a disordered phase in the lower half are present.

It is noteworthy that annealing of the sample from 80 K to room temperature leads to the same morphologies as direct deposition of TBPB at room temperature. Evaluating the center-to-center distances within these patterns confirm non-covalently interlinked molecules. Moreover, the structures are stabilized by an attractive electrostatic interaction similar to the low temperature phases of Cu(111) and Ag(111). The fact of coexisting phases indicates the relative weakness and topological versatility of the halogen-halogen interaction. The experimentally observed bonding schemes on different substrates with different crystallographic orientations and temperature are presented in Figure 5.15 (e).

As obvious from the experimental results, the choice of surface plays a major role for the formation process. In the following, the surface reactivity is explained in terms of its geometrical and electronic effects, where the former is related to the substrate orientation and the latter to the material.[256] In case of an ideal fcc (111) surface, a surface atom is coordinated by 9 atoms, while for an ideal fcc (110) surface the coordination number is 7. In fact, the (110)

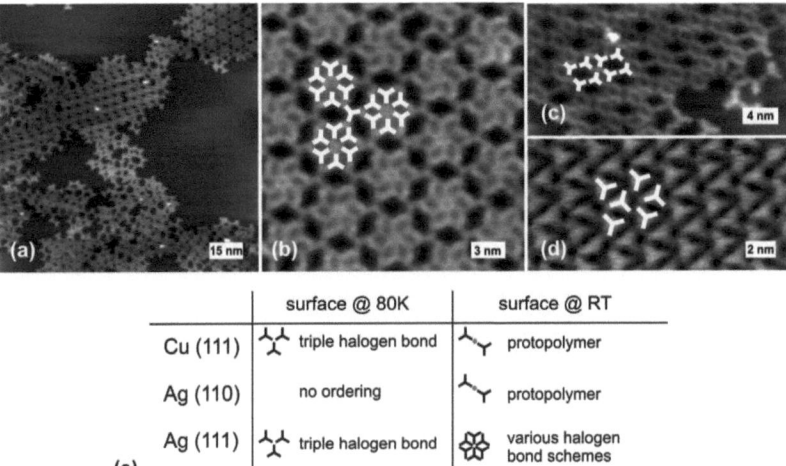

Figure 5.15: STM topographs of different phases of self-assembled TBPB molecules on Ag(111) at room temperature, while the images were acquired at 80 K for drift stability. (a) Overview image of two different coexisting phases, *i.e.* a row structure on the upper center part, a hexagonal flower structure in the upper right part, and a disordered phase in the lower half. (b) Close-up of the flower structure with tentative model overlaid. (c) Zoom-in image of the row structure. (d) Close-up of the densely packed structure. (e) Summary of the experimentally observed bonding schemes taking the substrate material and the crystallographic orientation as well as the temperature into account. The red dots represent adatoms of the particular surface.[232]

surface offers a large area of low coordinated surface atoms and, thus, catalytic activity is promoted. In general, the coordination number can directly influence the energy of the d band center and therewith the reactivity of the respective sites.[257] Furthermore, the adsorption of an aromatic molecule, for instance TBPB, on a transition metal results in a modification of the electronic structure (see section 4.3). Thereby, the homolysis of the C-Br bond is facilitated by a thermally activated charge transfer process from the newly occupied π^* into σ^* orbitals, which are anti-bonding with respect to the C-bromine bond. Thus, these bonds are destabilized followed by a dissociation of the C-Br bond. Prerequisite for this reaction, besides the temperature, is a sufficiently strong interaction strength of the molecules with the surface, which is fulfilled for Cu(111) but not for Ag(111).

5.5.2 2D COFs under ambient conditions[‡‡]

In order to reduce the effort of ultra-high vacuum experiments, the preparation protocols for the generation of COFs[156,207] were transferred to ambient conditions. Therefore, an Ullmann-type reaction was applied to halogenated monomers, *i.e.* 1,3,5-tris(4-iodophenyl)benzene (TIPB), on a metal catalyst.[259] Under ambient conditions, the choice of a metal substrate was restricted to Au(111) due to a lack of preparation methods for other metals. Moreover, the gold surface is well known to act as both the catalyst for the homolysis and the template for the resulting structures.[260]

Figure 5.16 (a) and (b) show STM topographs of TIPB molecules on Au(111) dissolved in nonanoic acid (9A) at room temperature. The resulting structures on the Au(111) surface depend on the initial concentration of the solution. For the low concentration pattern (0.02 mmol L^{-1}; Figure 5.16 (a)) a monolayer coverage emerged containing some small differently oriented domains of TIPB molecules. The angle between the differently oriented areas is a multiple of 30°, which is due to the substrate symmetry. In addition, some bright trigonal features are present, as shown in the inset on the lower right. These were identified as TIPB molecules adsorbed in a second layer by their perfect fit between the experimentally determined size and the optimized geometry of intact TIPB molecules with an iodine-iodine distance of approximately 1.4 nm.

For higher concentrations (0.04 mmol L^{-1}; Figure 5.16 (b)) a higher density of TIPB molecules within the second layer was observed. The bright features terminating the monomer were assigned to the peripheral iodine atoms. Interestingly, the arrangement of molecular pairs of TIPB in the second layer on Au(111) with those of the directly adsorbed TIPB monolayer on non-reactive graphite (see Figure 5.16 (c)) appears quite similar. Here, three or four iodine atoms are in close adjacency.

To trigger the homolysis of the halogen atoms and to start the interlinking of the TIPB molecules, the Au(111) substrate was heated to 100 °C and 5 µL of the solution c = 0.80 mmol L^{-1} was deposited on the surface. After 120 s on

---
[‡‡]submitted [258]

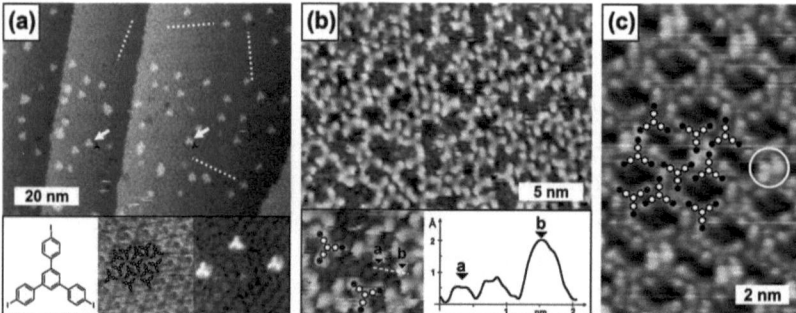

Figure 5.16: STM topographs of TIPB acquired in 9A solution on different surfaces: (a), (b) on Au(111) and (c) on graphite(001). (a) A TIPB concentration of $0.02\,\text{mmol}\,\text{L}^{-1}$ results in small ordered domains. Individually adsorbed TIPB molecules in the second layer are highlighted by white arrows and differently oriented domains in the first layer are indicated by white dotted lines. The insets depict the chemical structures of TIPB, a domain of TIPB molecules directly adsorbed on the Au(111) surface, and a zoom-in of second layer TIPB. (b) STM image showing the second layer with increased TIPB density, when the concentration was higher ($0.04\,\text{mmol}\,\text{L}^{-1}$). The inset shows a zoom-in, where some features of the first layer are still visible and terminating iodine atoms are highlighted by white dots. The line-profile confirms the formation of a bilayer system. (c) Self-assembled structure of TIPB with saturated solution on graphite. A molecular overlay indicates the tentative arrangement within the well-ordered monolayer. STM tunneling parameter: (a) $V_T = 0.26\,\text{V}$, $I_T = 12\,\text{pA}$; (b) $V_T = -0.62\,\text{V}$, $I_T = 61\,\text{pA}$; (c) $V_T = 0.74\,\text{V}$, $I_T = 46\,\text{pA}$.[258]

the hot plate and subsequent cooling, the networks were characterized by STM. Figure 5.17 (a) shows various aggregates of covalently interlinked monomers such as one-dimensional chains, open rings, closed pentagons, hexagons, heptagons, as well as more extended and irregular networks. A detailed analysis of the experimentally determined aryl-aryl bond length of 1.3 nm is in excellent agreement with density functional theory[261] and UHV experiments of topologically similar networks.[262] However, the formation of aggregates with up to 25 monomers yields lateral dimensions of up to 10 nm which represents a major step forward from previously reported coupling processes at the liquid-solid interface.

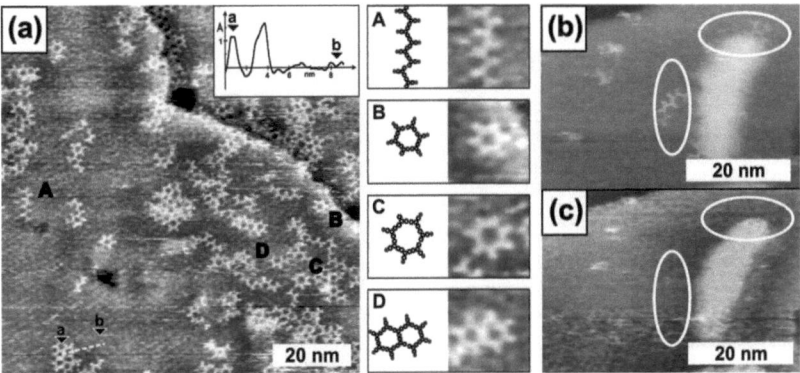

Figure 5.17: STM topograph of covalently interlinked TIPB monomers. The sample was prepared by deposition of 5 µL of TIPB solution in 9A (c = 0.80 mmol L^{-1}) onto Au(111) held at 100 °C for 120 s. The covalent aggregates are adsorbed on top of a first monolayer. Close-ups (A-D) of frequently encountered covalent aggregates are shown on the middle: (A) one-dimensional chains, (B) closed hexagons (C) closed heptagons, (D) more extended structures as merged rings; besides that more extended irregular structures and open rings were frequently observed. (b) and (c) Subsequently recorded STM images revealing the detachment process of covalent aggregates. Features are highlighted by white circles. STM tunneling parameter: (a) V_T = −0.752 V, I_T = 63 pA; (b) and (c) V_T = −0.624 V, I_T = 58 pA.[258]

A closer look at the covalently linked structures reveals that some legs of the monomers are often still terminated by iodine. This might be a possible reason for premature termination of the polymerization. Interestingly, the aggregates are distributed all over the substrate terraces and not restricted to step-edges. Moreover, changing the solvent to a shorter fatty acid such as heptanoic acid results in similar topologies. These covalent aggregates are not directly adsorbed on the Au(111) surface, but on top of a first monolayer. The second layer is rather weakly adsorbed as demonstrated by occasional observation of their detachment during scanning (see Figure 5.17 (b) and (c)).

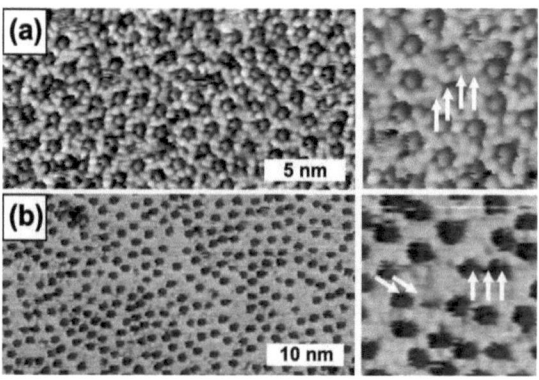

Figure 5.18: Deposition of TIPB onto Au(111) at a surface temperature of 100 °C either for 120 s (a) or 120 min (b) with a concentration of 0.80 mmol L^{-1}. The STM topographs reveal monolayer consisting of a mixture of mostly iodine atoms and partially dehalogenated TIPB monomers. On the right, zoom-ins are shown. White arrows highlight spherical protrusions that are assigned to chemisorbed iodine atoms. STM tunneling parameter: (a) $V_T = -0.348$ V, $I_T = 12$ pA; (b) $V_T = -0.359$ V, $I_T = 35$ pA.[258]

A detailed analysis of the first monolayer reveals a disordered mixture of adsorbed iodine and monomers, possibly partially dehalogenated. Figure 5.18 (a) and (b) depict the results, that were obtained when the sample was annealed either for 120 s or 120 min at 100 °C. It is clearly visible from the STM topographs, that shorter annealing times favor the formation of nanopores with single bright features at the center. For longer annealing times the area density of iodine is increased. All show the same typical spacing of 0.5 nm. This distance can be matched to the characteristic lattice parameter of 0.5 nm for a ($\sqrt{3} \times \sqrt{3}$)R30° iodine superstructure on Au(111).[132] In a control experiment, TIPB solution was applied to iodine terminated Au(111) surfaces, kept at 100 °C. As anticipated, no covalent structures were observed due to poisoning effects of the adsorbed iodine atoms.

To further the understanding of the progression of the halogenation, XPS measurements were performed. Therefore, TIPB was drop-cast on a Au(111) surface, held at room temperature. In order to stop further TIPB adsorption, the sample was rinsed with ethanol for 60 s before it was transferred into the UHV system. The first XPS signatures were acquired from I(3d) core levels. For unreacted TIPB compound on Au(111) the energies of the spin-orbit doublet for the I_{Phenyl} occurs at 620.4 eV (I-$3d_{5/2}$) and 632.0 eV (I-$3d_{3/2}$), respectively. These values are in perfect agreement to comparable iodinated compounds, *i.e.* the iodophenyl isocyanate I-$3d_{5/2}$ peak at 620.47 eV. However, the characteristic binding energies for iodine-gold are clearly distinguishable, *i.e.* I-$3d_{3/2}$ 619.0 eV. Figure 5.19 shows the XPS spectra of TIPB on Au(111), recorded at room temperature (blue curve), 100 °C (green curve), and 150 °C (red curve). As obvious from the presented spectra, some spontaneous dehalogenation even occurs at room temperature. Annealing the sample to higher temperatures, the ratio of the two chemically distinct iodine species changes. At temperatures of 150 °C unreacted TIPB molecules disappear, indicating a complete homolysis. Nevertheless, the total amount of iodine on the surface remains constant, as verified by the constant iodine/Au ratio of approximately 2.0 atom-% based on an evaluation of the total peak areas.

It is known from previous UHV studies of the bromine analogs (TBPB) that thermal activation, *i.e.* 140 °C to 180 °C, is essential to cleave the halogen substituents.[263] Russell *et al.* reported on the formation of TBPB dimers under ambient conditions and determined activation temperatures of 200 °C.[102] Due to the lower bond dissociation energy of C-I bonds (2.84 eV in iodobenzene) in contrast to the C-Br bonds (3.49 eV in bromobenzene)[264] lower surface temperatures are required to trigger the dehalogenation. Moreover, this effect is supported by a rather strong gold-iodine bond.

For the generation of such covalently interlinked flakes, the following reaction scheme is proposed: The dehalogenation rate of TIPB molecules on the catalytic Au(111) surface is significantly enhanced by increasing the sample temperatures. After the homolysis had occurred on the annealed surface, the radicals recombined and formed new covalently interlinked aggregates. Interestingly, these

Two dimensional molecular structures

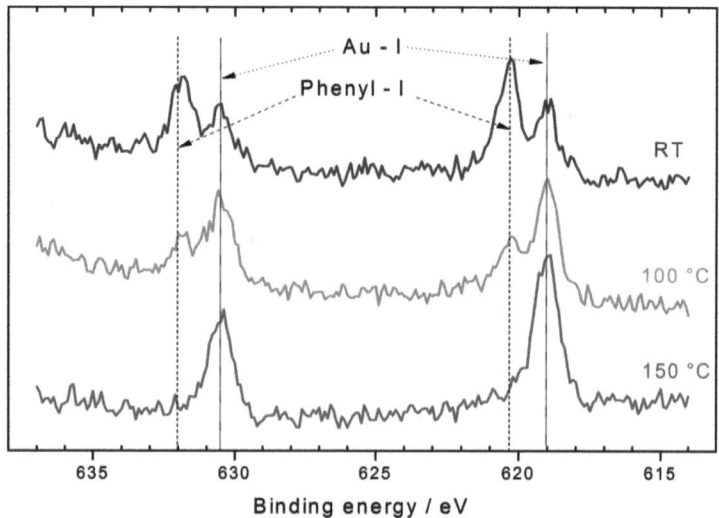

Figure 5.19: XPS spectra of TIPB deposited onto Au(111) under ambient conditions at various temperatures. The curves are vertical offset for clarity. The blue curve shows the data recorded at room temperature, whereas the green curve was obtained after heating the sample for 10 min up to 100 °C, and the red curve after further heating of the same sample for 10 min up to 150 °C.[258]

organic networks were adsorbed on a layer of split-off iodine atoms on top of the Au(111) surface. The iodine layer suppresses further homolysis, but becomes the supporting template for the covalent networks.

CHAPTER 6

Summary and Outlook

This thesis shed light on the processes of the formation of molecular structures, namely the creation of self-assembled 2D networks, 2D metal-organic frameworks (MOFs), and 2D covalent organic frameworks (COFs). These patterns emerge from a complex interplay between adsorbate-substrate as well as adsorbate-adsorbate interactions and are typically influenced by a variety of external parameters. In order to investigate and characterize the dimensions at this length scale, sophisticated experimental setups and devices are essential. A brief introduction to scanning tunneling microscopy, the fundamental analysis device used in this thesis, the used materials and methods, and a description of the fundamental interactions at the nanoscale were given.

The incorporation dynamics of coronene guest molecules into pre-existent 2D host networks from trimesic acid (TMA) or the larger analogous benzenetribenzoic acid (BTB), was studied. When adding some guest-containing solution to the TMA system forming a stable honeycomb host network on the surface, transient intermediate states were identified. In contrast, an immediate incorporation of the same guest in the significantly larger BTB pores was observed within the temporal resolution of the STM experiment, $i.e.$ < 0.5 s. Furthermore, a solvent dependence of the incorporation dynamics has been discovered for the TMA network. For the shorter heptanoic acid intermediate states have been observed only very rarely. For the incorporation into TMA pores, molecular mechanics suggests an additional stabilization of the COR molecules within the pores by 18 newly formed hydrogen bonds between COR and the pore wall. Due to the fact, that the pores are not empty, but occupied by solvent molecules, these molecules need to be displaced for the COR to enter the void. This process is much faster for heptanoic acid than for nonanoic acid and also much faster for BTB pores than for geometrically matched TMA pores due to differences in pore size.[92]

Furthermore, the synthesis of metal-organic frameworks (MOFs) based on aromatic trithiols was investigated under ultra-high vacuum conditions. Therefore, 1,3,5-tris(4-mercaptophenyl)benzene (TMB) molecules were deposited either on Cu(111) or Ag(111) substrates kept at room temperature. The precursor structures on both surfaces were composed of densely packed trigonal structures, where the TMB molecules were already deprotonated. Annealing the Cu(111) to 150 °C yields metal-coordination bonds between the thiolates and either one or two coordinating copper adatoms, depending on the initial surface coverage. In contrast to Cu(111), annealing the TMB on the Ag(111) sample at 300 °C results in irregular structures, where covalent disulfur bridges were identified. These differences are attributed to the lower affinity of silver adatoms to form metal-coordination bonds with thiolates as compared to copper adatoms. In general, thiol groups coordinated via copper adatoms are suited for application in molecular electronics due to their electronic conjugation, their high stability, and the way of creating them by bottom-up strategies.[238]

A further issue addressed in this work were covalent organic frameworks (COFs). These studies were carried out under ultra-high vacuum conditions as well as under ambient conditions on reactive metal surfaces. Under both environmental conditions, the focus was on covalent coupling through radical addition reactions of appropriate monomers, *i.e.* halogenated aromatic molecules such as 1,3,5-tris(4-bromophenyl)benzene (TBPB) and 1,3,5-tris(4-iodophenyl)benzene (TIPB). Here, a thermal activation energy for the scission of the carbon-halogen bonds is essential besides the reactive surface itself. In the case of ultra-high vacuum experiments TBPB was deposited on to different surfaces, *i.e.* Cu(111), Ag(111), and Ag(110), kept at different temperatures. The homolysis of the C-Br bond and subsequent formation of proto-polymers only occur for room temperature deposition of TBPB on Cu(111) but not on Ag(111). This clearly underpins the material dependency of the catalytic capability. By further annealing, these proto-polymers can be converted to 2D covalent networks. However, for Ag(110) the debromination reaction did not take place indicating a dependency on the crystallographic orientation. Moreover, deposition onto Cu(111) and Ag(110) kept at low temperature (80 K) lead to triply halogen bonded networks. For Ag(110), only unordered adsorbed molecules were present at this temperature. These findings underline the necessity of thermal activation.[232,233]

It was shown that the polymerization of the triply iodinated monomer TIPB can be initiated by drop-casting on a pre-heated Au(111) surface. In contrast to similar experiments under ambient conditions with TBPB, the brominated analogue of TIPB, more extended covalent structures that consist of up to 25 monomeric units were obtained, while the brominated monomer only yielded dimers. So the present system is a further example to corroborate the enhanced reactivity of iodinated precursors for the proposed polymerization reaction. Interestingly, most of the covalent structures were found on top of a first monolayer. This first monolayer is mostly comprised of iodine, but also partially dehalogenated TIPB monomers were resolved. Since the lifting of the polymerized aggregates from the metal substrate is unprecedented from UHV experiments, the environmental conditions might play the decisive role. Moreover, it was shown that deliberately iodine terminated Au(111) surfaces lack the catalytical activity to promote the initiating dehalogenation reaction. Thus, progressive iodine adsorption terminates the polymerization reaction. On the other hand the loosely bound covalent aggregates on top of iodine are weakly adsorbed and promising for lift-off and transfer to other substrates. In summary, iodinated compounds are ideal candidates for polymerization under ambient conditions. The enhanced reactivity allows to operate with relatively inert surfaces such as gold at the price of relatively fast iodine poisoning and deactivation of the catalytic surface.

In general, self-assembled monolayers (SAMs) bear great technological potential and are offering a promising strategy for mass fabrication of complex molecular systems. A specially tailored combination of precursor molecules enables the formation of more complex chemical products and thus offers new functionalities. Also the co-adsorption of specific guest molecules onto different surfaces or pre-existing host-networks opens up new perspectives. In this context,

Summary and Outlook

appropriate theoretical calculations can be used to design new materials based on known reaction schemes.

To date, the generation of large areas of 2D covalent networks which are applicable for novel devices has not been realized. A promising strategy is the conversion of defect-free self-assembled structures (10^{14} molecules/cm^2) into stable 2D covalent networks. In order to tune the electronic band structure and therewith the properties of these networks, either appropriate building blocks have to be used for the fabrication or further organic molecules have to be adsorbed on these networks.[236]

In future, one of the main challenges will be the interfacing of such nanostructures to the macroscopic world. Research has to be addressed to both contacting and transport properties in order to unleash the full potential of the newly accessible 2D nanostructures.

CHAPTER 7

Publications

In the following, the publications including the supplementary information are re-printed with the permission of the respective journals.

7.1 Surface mediated synthesis of 2D covalent organic frameworks: 1,3,5-tris(4-bromophenyl)benzene on graphite(001), Cu(111), and Ag(110) . . 116

7.2 Material- and orientation-dependent reactivity for heterogeneously catalyzed carbon-bromine bond homolysis 120

7.3 A combined ion-sputtering and electron-beam annealing device for the *in-vacuo* post preparation of scanning probes 127

7.4 Extended two-dimensional metal-organic frameworks based on thiolate-copper coordination bonds 132

7.5 Incorporation dynamics of molecular guests into two-dimensional supramolecular host networks at the liquid-solid interface 140

7.6 Solution preparation of two dimensional covalently linked networks by polymerization of 1,3,5-tri(4-iodophenyl)benzene on Au(111) 150

7.1 Surface mediated synthesis of 2D covalent organic frameworks: 1,3,5-tris(4-bromophenyl)benzene on graphite(001), Cu(111), and Ag(110)

Rico Gutzler, Hermann Walch, Georg Eder, Stephan Kloft,
Wolfgang M. Heckl, and Markus Lackinger

Chem. Commun., **2009**, 4456–4458

http://dx.doi.org/10.1039/b906836h

Reproduced by permission of The Royal Society of Chemistry (RSC)

Surface mediated synthesis of 2D covalent organic frameworks: 1,3,5-tris(4-bromophenyl)benzene on graphite(001), Cu(111), and Ag(110)†

Rico Gutzler,[*a] Hermann Walch,[a] Georg Eder,[a] Stephan Kloft,[a] Wolfgang M. Heckl[ab] and Markus Lackinger[*a]

Received (in Cambridge, UK) 7th April 2009, Accepted 28th May 2009
First published as an Advance Article on the web 12th June 2009
DOI: 10.1039/b906836h

The on surface synthesis of a two-dimensional (2D) covalent organic framework from a halogenated aromatic monomer under ultra-high vacuum conditions is shown to be dependent on the choice of substrate.

The synthesis of 2D covalent organic frameworks (COF) on surfaces has recently gained much attention.[1] Commonly, these novel polymers are built by sublimation of appropriate monomers onto metallic substrates under ultra-high vacuum (UHV) conditions and subsequent annealing.[2–6] Many other studies investigated the formation of covalent structures from smaller building blocks and demonstrated the importance of the substrate both for the confinement of molecular motion in two dimensions and as a catalyst for activation.[7–10]

Here, we report on the reticular synthesis of 2D COFs built up from conjugated subunits (phenyl rings) only. The halogen substituted polyaromatic monomer is thermally sublimed onto various substrates under UHV conditions at room temperature. Without providing additional activation energy, the formation of radicals is observed on Cu(111) and Ag(110). Deposition of the same compound on graphite(001) results in non-covalent self-assembly of well ordered networks stabilized by halogen–hydrogen bonds. These results demonstrate the decisive role of the substrate for homolysis of covalent carbon–halogen bonds at room temperature and subsequent association of radicals.

Our strategy for the synthesis of 2D COFs consists of the deposition of a suitable organic compound and its subsequent substrate mediated homolysis. Intermolecular colligation occurs through radical addition at elevated temperatures. For formation of 2D open-pore networks from a single kind of molecule the building block must at least be a triradical, *i.e.* exhibit three potential binding sites. Comparatively weak carbon–halogen bonds are well suited as predetermined breaking points and from solution chemistry Cu catalysts are known to facilitate homolysis.[11] The molecule of choice is 1,3,5-tris(4-bromophenyl)benzene (TBB, *cf.* Fig. 1a). TBB is an appropriate candidate because the homolytic bond dissociation energy of its C–Br bonds (3.2 eV) is substantially lower than that of the C–C link between phenyl rings (4.8 eV).[12,13]

Evaporation of a TBB monolayer on graphite(001) results in an ordered structure (*cf.* Fig. 1b and c). Although no submolecular resolution is achieved, the long range order and the mutual arrangement of molecules within the monolayer indicate non-covalent self-assembly of intact TBB molecules.

In agreement with molecular mechanics simulations, molecules arrange in chains of dimers which are stabilized by Br···H–C hydrogen bonds. Similar binding motifs have been reported for comparable systems[14–16] and also stabilize the bulk structure of TBB.[17] The contrast modulation, *i.e.* every other row appears brighter, is caused by different adsorption sites on the graphite lattice. Annealing of the graphite sample to ~320 °C for 10 min results in complete desorption of the TBB monolayer, thereby demonstrating the inferior stability of this non-covalent network.

Open-pore networks can clearly be identified in the STM topographs obtained for submonolayer coverage of TBB on Cu(111) (Fig. 2) and Ag(110) (*cf.* Fig. S1, ESI†). These

Fig. 1 (a) Structure of 1,3,5-tris(4-bromophenyl)benzene (TBB). (b) STM topograph of a TBB monolayer on graphite (V = 1.9 V, I = 76 pA). Unit cell is indicated by blue lines (a = 3.4 nm, b = 2.5 nm, α = 44°) and accommodates four molecules, symbolized by blue tripods. (c) Mesh-averaged image (4.1 × 5.0 nm^2) of (b).

[a] *Department of Earth and Environmental Sciences and Center for NanoScience (CeNS), Ludwig-Maximilians-University, Theresienstrasse 41, 80333, Munich, Germany.*
E-mail: rico.gutzler@lrz.uni-muenchen.de, markus@lackinger.org;
Tel: +49 89 21804217
[b] *Deutsches Museum, Museumsinsel 1, 80538, Munich, Germany*
† Electronic supplementary information (ESI) available: Experimental details, calculations, thermogravimetric analysis, additional STM data, and UV/Vis spectra. See DOI: 10.1039/b906836h

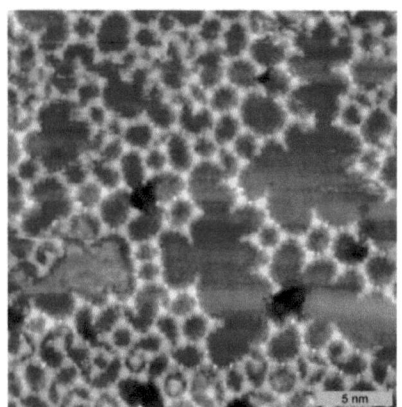

Fig. 2 TBB protopolymer on Cu(111) before annealing ($V = -1.85$ V, $I = 248$ pA). Spherical protrusions between radicals are clearly observable.

networks are composed of polygons, predominantly hexagons and pentagons but also heptagons, octagons and other polygons. The experimentally determined values for the center-to-center distance between adjacent molecules for room temperature deposition are 1.49 nm ± 0.10 nm for Cu(111) and 1.57 nm ± 0.06 nm for Ag(110) and are equal for both substrates within the error margin. These values are somewhat larger than anticipated for a covalent C–C link between adjacent molecules. In the topograph of the TBB network on Cu(111) (Fig. 2), bright protrusions can clearly be discerned between adjacent molecules. These spherical features are attributed to single copper atoms which coordinate two or occasionally three radicals. Similar systems based on halogenated benzene derivatives were reported to assemble in a first step into so-called protopolymers where two radicals are linked via a metal atom.[7,18] This preceding formation of a metal coordination complex between on surface-generated radicals and substrate atoms is also likely to be observed here. Coordinating atoms could not be resolved on Ag(110), but the spacing clearly indicates formation of protopolymers as well. In this respect, no difference between the two metal substrates was found, although (110) surfaces exhibit a pronounced anisotropy. In most cases the network structures are attached to step edges, hence it is probable that growth is initiated by attachment of a free radical to a step edge (Fig. S1, ESI†).

In order to verify whether protopolymers can eventually be converted into COFs, annealing experiments have been carried out with the Cu(111) surface. Fig. 3a depicts an STM topograph of a tempered sample. After annealing to 300 °C the distance between two interconnected molecules is reduced to 1.24 nm ± 0.06 nm (Fig. 3a). Accordingly, the linescan in Fig. 3b yields a size of 2.2 nm for each hexagon. The STM contrast of the network after annealing at 300 °C (Fig. 3a) is very different from the network before annealing (Fig. 2). The bright features in between two molecules are absent in the

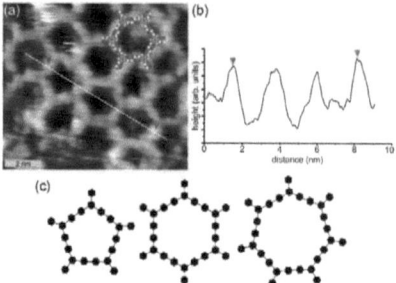

Fig. 3 TBB covalent organic framework on Cu(111) (a) evaporation temperature of 140 °C for 8 minutes ($V = +3.2$ V, $I = 233$ pA) and subsequent annealing to 300 °C. A scaled hexagonal ring is overlaid. (b) Linescan as indicated in (a) across three rings yielding a center-to-center distance of 2.2 nm for a single ring. (c) Principally encountered binding motifs: pentagon, hexagon and heptagon. Bonds between phenyl rings are distorted by $\Delta\alpha$.

post-annealing topograph, indicating a physical change in the monolayer. More importantly, the reduced size (both the lattice constant and the spacing of adjacent molecules) upon annealing indicates the transition from a protopolymer to a 2D COF. Lipton-Duffin et al. found for polymerization experiments with p-diiodobenzene that the phenyl–phenyl spacing is reduced by 0.12 nm upon transition from a protopolymer to an actual covalent linkage.[7] This value is consistent with our observations. The necessity of an additional annealing step to finally induce covalent linkage was also experienced by other groups for comparable systems.[7,19]

The distance between two monomers in the post-annealing network is also in accordance with both DFT and molecular mechanics simulations of an infinite layer of TBB molecules covalently linked at the 4′-position (1.28 nm, experimental: 1.24 nm ± 0.06 nm). A hypothetical hexagonal network based on intact TBB subunits pointing head-to-head with their bromine atoms would necessarily result in a considerably larger center-to-center distance of ~1.75 nm as estimated by molecular mechanics calculations and can thus be excluded.

The high defect density, that is the frequent occurrence of polygons different from hexagons, can be explained by considering the energy necessary to bend one bond between two phenyl groups: due to the threefold symmetry of TBB, a hexagonal ring comprising six molecules would yield the lowest-energy geometry because all newly formed links exhibited an ideal bonding angle of 180°. All other polygons experience slightly higher stress due to distortion of the bond angle. Assuming regularity and rigid phenyl rings, each bond between two phenyl rings in the polygon is distorted from its optimal equilibrium angle on average by $\Delta\alpha = 60°(6 - N)/(3N)$, where N is the number of molecules in the ring (Fig. 3c). The change $\Delta\alpha$ is small and accounts to 4°, 0°, and −2.9° for pentagon, hexagon, and heptagon, respectively, yielding only a small additional energy contribution. This small deviation from the equilibrium geometry is responsible for the high number of polygons different from hexagons. Since colligation of free radicals is virtually barrier free and only diffusion limited, kinetic effects can result in suboptimal binding geometries where the binding angle can deviate from 180°. This leads to a reduced order and high defect densities.

Interestingly, the chemical activity of the substrate has a major contribution to cleavage of the C–Br bond. The activation of TBB molecules, that is generation of triple radicals by cleavage of all three covalent C–Br bonds, requires a metallic substrate. In general, the binding energy of halides chemisorbed to metal surfaces is particularly strong and was found to be in the order of 1.5 eV. Thus, a low free energy of the final state certainly promotes homolytic fission of C–Br bonds in TBB. Physisorption of halides on graphite would render homolysis strongly endothermic. In some STM topographs on metal surfaces, circular features appear after deposition of TBB and can be attributed to adsorbed Br atoms.

For the likewise covalently linked structures prepared from Br substituted tetraphenyl porphyrins and dibromoterfluorene on catalytically less active Au(111) surfaces, significantly higher temperature thresholds of 315 °C and 250 °C for thermal activation are reported.[2,20] The first compound is much heavier than TBB and the temperature for bond cleavage is in the regime of the sublimation temperature, thus activation can already occur in the crucible. In contrast, for TBB thermal evaporation of non-activated species is easily possible. The distinct substrate dependence clearly demonstrates that homolysis takes place on the surface and not in the crucible as observed for other systems.[2,6] Furthermore, UV/Vis spectroscopy independently confirms that the TBB molecules are intact prior to sorption on the surface and do not dissociate at the sublimation temperatures of 140–160 °C (cf. ESI†).

Thermal stability of the COFs on Cu(111) has been verified by further annealing experiments and subsequent STM characterization. Annealing of the Cu(111) sample at 400 °C caused degradation of the networks and STM images no longer exhibit open-pore structures (see Fig. S2, ESI†, for a degraded network). Thermogravimetric analysis of pure TBB reveals an onset for decomposition at a temperature of around 250 °C (cf. ESI†), which is somewhat lower than for the COF. Its higher thermal stability can be attributed to the absence of comparatively weak C–Br bonds in the monolayer, strong intermolecular bonds, and interaction with the substrate.

In this work, we demonstrate the formation of substrate supported 2D COFs by addition of on surface-generated triple radicals. Experiments on different surfaces unveil the important role of the substrate for the main activation step, homolytic fission of C–Br bonds. On metal substrates where the split-off Br-atoms are stabilized by strong chemisorption, homolysis takes place without providing additional activation energy. Chemically inert graphite surfaces cannot promote homolysis, thus cannot initiate formation of covalent bonds. However, due to preceding formation of a protopolymer through metal coordination of radicals, an additional thermal activation is required to transfer the networks eventually into COFs. In this respect it would be highly interesting to find either a system, a method, or conditions where on one hand the substrate is catalytically effective for homolysis but on the other hand formation of protopolymers is suppressed. The immediate formation of covalent bonds would definitely change the association kinetics and will thus possibly also influence the ordering.

Financial support by Deutsche Forschungsgemeinschaft (SFB 486) and Nanosystems Initiative Munich (NIM) is gratefully acknowledged. Georg Eder acknowledges support by the Hanns-Seidel-Stiftung.

Notes and references

1 D. F. Perepichka and F. Rosei, *Science*, 2009, **323**, 216–217.
2 L. Grill, M. Dyer, L. Lafferentz, M. Persson, M. V. Peters and S. Hecht, *Nat. Nanotechnol.*, 2007, **2**, 687–691.
3 M. Treier, N. V. Richardson and R. Fasel, *J. Am. Chem. Soc.*, 2008, **130**, 14054–14055.
4 M. In't Veld, P. Iavicoli, S. Haq, D. B. Amabilino and R. Raval, *Chem. Commun.*, 2008, 1536–1538.
5 S. Weigelt, C. Busse, C. Bombis, M. M. Knudsen, K. V. Gothelf, E. Laegsgaard, F. Besenbacher and T. R. Linderoth, *Angew. Chem., Int. Ed.*, 2008, **47**, 4406–4410.
6 N. A. A. Zwaneveld, R. Pawlak, M. Abel, D. Catalin, D. Gigmes, D. Bertin and L. Porte, *J. Am. Chem. Soc.*, 2008, **130**, 6678–6679.
7 J. A. Lipton-Duffin, O. Ivasenko, D. F. Perepichka and F. Rosei, *Small*, 2009, **5**, 592–597.
8 M. Matena, T. Riehm, M. Stöhr, T. A. Jung and L. H. Gade, *Angew. Chem., Int. Ed.*, 2008, **47**, 2414–2417.
9 S. Weigelt, C. Bombis, C. Busse, M. M. Knudsen, K. V. Gothelf, E. Laegsgaard, F. Besenbacher and T. R. Linderoth, *ACS Nano*, 2008, **2**, 651–660.
10 S. Weigelt, J. Schnadt, A. K. Tuxen, F. Masini, C. Bombis, C. Busse, C. Isvoranu, E. Ataman, E. Laegsgaard, F. Besenbacher and T. R. Linderoth, *J. Am. Chem. Soc.*, 2008, **130**, 5388–5389.
11 S. V. Ley and A. W. Thomas, *Angew. Chem., Int. Ed.*, 2003, **42**, 5400–5449.
12 R. J. Kominar, M. J. Krech and S. J. W. Price, *Can. J. Chem.*, 1978, **56**, 1589–1592.
13 M. Szwarc, *Nature*, 1948, **161**, 890–891.
14 G. R. Desiraju and R. Parthasarathy, *J. Am. Chem. Soc.*, 1989, **111**, 8725–8726.
15 H. F. Lieberman, R. J. Davey and D. M. T. Newsham, *Chem. Mater.*, 2000, **12**, 490–494.
16 O. Navon, J. Bernstein and V. Khodorkovsky, *Angew. Chem., Int. Ed. Engl.*, 1997, **36**, 601–603.
17 L. M. C. Beltran, C. Cui, D. H. Leung, J. Xu and F. J. Hollander, *Acta Crystallogr., Sect. E*, 2002, **58**, O782–O783.
18 G. S. McCarty and P. S. Weiss, *J. Am. Chem. Soc.*, 2004, **126**, 16772–16776.
19 M. Xi and B. E. Bent, *Surf. Sci.*, 1992, **278**, 19–32.
20 L. Lafferentz, F. Ample, H. Yu, S. Hecht, C. Joachim and L. Grill, *Science*, 2009, **323**, 1193–1197.

7.2 Material- and orientation-dependent reactivity for heterogeneously catalyzed carbon-bromine bond homolysis

Hermann Walch, Rico Gutzler, Thomas Sirtl, Georg Eder, and Markus Lackinger

J. Phys. Chem. C, **2010**, 114, 12604–12609

http://dx.doi.org/10.1021/jp102704q

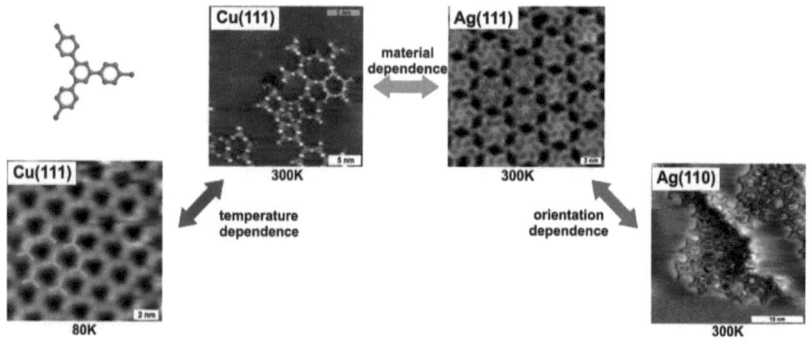

Reproduced with permission from [232]. Copyright 2010 American Chemical Society

Material- and Orientation-Dependent Reactivity for Heterogeneously Catalyzed Carbon−Bromine Bond Homolysis

Hermann Walch,*,† Rico Gutzler,† Thomas Sirtl,† Georg Eder,† and Markus Lackinger*,†,‡

Department for Earth and Environmental Sciences and Center for NanoScience (CeNS), Ludwig-Maximilians-University, Theresienstrasse 41, 80333 Munich, Germany, and Deutsches Museum, Museumsinsel 1, 80538 Munich, Germany

Received: March 25, 2010; Revised Manuscript Received: June 18, 2010

Adsorption of the brominated aromatic molecule 1,3,5-tris(4-bromophenyl)benzene on different metallic substrates, namely Cu(111), Ag(111), and Ag(110), has been studied by variable-temperature scanning tunneling microscopy (STM). Depending on substrate temperature, material, and crystallographic orientation, a surface-catalyzed dehalogenation reaction is observed. Deposition onto the catalytically more active substrates Cu(111) and Ag(110) held at room temperature leads to cleavage of carbon−bromine bonds and subsequent formation of protopolymers, i.e., radical metal coordination complexes and networks. However, upon deposition on Ag(111) no such reaction has been observed. Instead, various self-assembled ordered structures emerged, all based on intact molecules. Also sublimation onto either substrate held at ∼80 K did not result in any dehalogenation, thereby exemplifying the necessity of thermal activation. The observed differences in catalytic activity are explained by a combination of electronic and geometric effects. A mechanism is proposed, where initial charge transfer from substrate to adsorbate, followed by subsequent intramolecular charge transfer, facilitates C−Br bond homolysis.

Introduction

Heterogeneous catalysis provides the basis for the economic synthesis of the majority of compounds produced worldwide and is thus of utmost importance for the chemical industry. In relation to its importance, however, the atomistic understanding of the underlying processes lags behind. The "surface science approach", which was introduced by Ertl, i.e., the use of atomically flat and clean single crystal surfaces under ultrahigh vacuum (UHV) conditions as model catalysts,[1,2] has stimulated a lot of effort in this field. Among other techniques, scanning tunneling microscopy (STM) has been taking a major role as a tool to reveal catalytic phenomena by high-resolution real space imaging,[3-6] in particular for the dissociative adsorption of molecules.[7,8]

Here we report on the heterogeneously catalyzed dehalogenation of the comparatively large aromatic compound 1,3,5-tris(4-bromophenyl)benzene (TBB, cf. Fig. 1 for structure) on coinage metal surfaces. TBB is also a well-suited candidate monomer for the synthesis of surface supported two-dimensional polymers. For the synthesis of two-dimensional polymers, different strategies are proposed to cleave the C−Br σ-bonds, an activation step which creates free radicals that can subsequently form covalent bonds through addition reactions. Recently, we could show that for the on-surface polymerization the substrate does not merely serve as support, but takes a vital chemical role.[9] An inert substrate like graphite(001) does not catalyze the surface-mediated homolysis and hence leaves the molecules intact upon physisorption, whereas on Cu(111) and Ag(110) the dehalogenation reaction readily occurs. Yet, instead of directly forming covalent intermolecular bonds, the on-surface generated radicals coordinate to surface supplied metal atoms in an intermediate reaction step. Thereby coordination complexes introduced as "protopolymers" by Weiss and co-workers are formed[10] through a surface-mediated reaction that has meanwhile also been observed for other systems.[11-13] The first reaction step is dissociative adsorption of halogenated aromatic species on a copper catalyst, a reaction scheme that resembles the coupling chemistry described by Ullmann in 1901.[14] Cleavage of carbon−halogen bonds followed by the formation of comparatively strong bonds[15] of the resulting radicals to copper atoms is also an intermediate step in the Ullmann coupling reaction. In the original Ullmann reaction the bidentate radical−copper complex is a short-lived reaction intermediate, while for the surface variant the radical−copper complexes are metastable at room temperature. Subsequent thermal annealing releases the coordinating copper atoms and induces covalent C−C coupling of the aromatic species. The initially split-off bromine species binds to the surface and thermally activated diffusion results in island formation at ∼600 K,[16] whereas desorption takes place at about 950 K.[17] In STM topographs adsorbed bromine atoms appear as depressions on Cu(111), which was explained by quenching of the surface state.[17] In accordance with these findings we also occasionally observe these depressions in the vicinity of protopolymers (see yellow arrow in Figure 1a), which we attribute to split-off bromine atoms, although we cannot unambiguously prove it.

In order to study the role of the metal support in more detail and shed light on the homolysis mechanism, further experiments were conducted on Ag(111), Ag(110), and Cu(111). The substrate temperature during deposition was introduced as an additional parameter. In this study, TBB was deposited on each substrate held at either room temperature or cooled down to ∼80 K. Subsequently, STM was applied to study the intermolecular bonding schemes and to identify products of a possible dehalogenation reaction.

---

* To whom correspondence should be addressed. E-mail: hermann.walch@physik.uni-muenchen.de (H.W.), markus@lackinger.org (M.L.).
† Ludwig-Maximilians-University.
‡ Deutsches Museum.

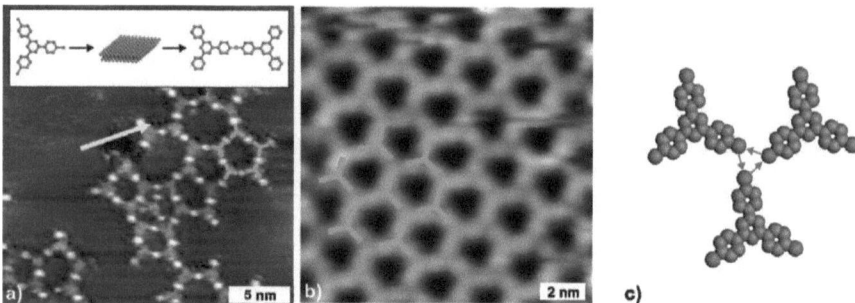

Figure 1. STM topographs of TBB deposited onto Cu(111) with the substrate held at (a) room temperature and (b) ~80 K, respectively. (a) Room-temperature deposition readily induces the dehalogenation reaction and subsequent formation of protopolymers ($U_T = 1.50$ V, $I_T = 85$ pA; inset, reaction scheme). (b) Deposition onto Cu(111) at ~80 K leads to noncovalent self-assembly of a highly ordered structure ($U_T = -1.98$ V, $I_T = 90$ pA). Due to their size and symmetry, the 3-fold bright features are assigned to single intact TBB molecules as shown in the overlay. (c) Tentative model of the intermolecular arrangement based on STM data. As a consequence of a nonspherical charge distribution around the halogen substituents an electrostatic stabilization known as a halogen–halogen bond becomes feasible.

Results and Discussion

Cu(111). Deposition of TBB onto Cu(111) at room temperature leads to the spontaneous formation of protopolymers, in accordance with previous experiments on brominated aromatic molecules.[11,12,17] Coordinating copper atoms are either extracted from terraces[18] or supplied by the free adatom gas that originates from a temperature-dependent condensation/evaporation equilibrium at step edges.[19] An STM topograph of TBB–protopolymer networks on Cu(111), i.e., radical–metal coordination complexes, is depicted in Figure 1a. Bright circular protrusions midst the triangular molecular units are clearly discernible, and readily identified as copper atoms. However, an unambiguous experimental indication for protopolymer formation is the center-to-center distance between interlinked molecules. In full agreement with the anticipated value for protopolymers, a distance of ~1.50 nm was found. The irregularity and high defect density of these networks is owed to both the pronounced reactivity of phenyl radicals and the low directionality of coordination bonds. By virtue of a postannealing step (up to 300 °C), it was possible to release the copper atoms and eventually convert metal-coordination bonds into covalent C–C interlinks. This is accompanied and proven by a ~0.25 nm decrease of the center-to-center distance of adjacent interconnected TBB molecules from ~1.50 nm to ~1.25 nm.[9,12]

In order to gain deeper insight into the dissociation mechanism, the present study also takes the influence of the substrate temperature during deposition into account. In the case of Cu(111), a prominent difference arises depending on the substrate temperature: Deposition of TBB onto Cu(111) held at ~80 K leads to the formation of highly ordered, virtually defect-free self-assembled structures which are comprised of intact molecules. A representative STM topograph and the corresponding structural model are depicted in Figure 1b,c. Single molecules are clearly resolved and appear as 3-fold symmetric features in accordance with the molecular structure. The structure is based on a hexagonal lattice with $a = 2.05 \pm 0.06$ nm and contains one molecule per unit cell. Both, the high degree of ordering and the unit cell dimensions substantiate the conclusion that molecules remain intact and self-assemble due to relatively weak noncovalent interactions. The halogen substituents cannot be distinguished from the aromatic backbone in the submolecular STM contrast, because the frontier molecular orbitals equally have contributions from the aromatic system and the peripheral halogen substituents, respectively. A comparable cyclic bonding pattern among three halogen atoms has previously been observed in bulk crystals of halogenated phenyls.[20] The underlying interaction is of electrostatic origin and attributed to a nonspherical charge distribution around the bromine substituents. Calculations of the electrostatic potential at the halogen atoms propose a positive cap opposite to the C–Br bond and a ring of negative potential around the bond axis.[21] A cyclic intermolecular arrangement as shown in Figure 1c thus optimizes electrostatic interactions and can be described as Coulombic "donor–acceptor" attraction.[20]

Warming up the well-ordered TBB layer on Cu(111) to room temperature also induces the formation of protopolymers, similar to those observed for room-temperature deposition. These experimental findings illustrate that dissociation of C–Br bonds on Cu(111) requires thermal activation and that the thermal energy supplied at 300 K is sufficient.

Ag(110). Regarding the dissociation of C–Br bonds upon room temperature deposition, Ag(110) shows a qualitatively similar behavior as Cu(111): formation of protopolymers was readily observed (Figure 2a). For low-temperature deposition, however, TBB molecules do not form ordered structures; instead, adsorption of isolated, apparently immobile single molecules has been observed, as illustrated in Figure 2b. This result is explained by a more corrugated surface potential on Ag(110) as compared to Cu(111). For face-centered cubic (fcc) metals, the potential energy landscape for adsorbates exhibits higher corrugation on (110) than on (111) surfaces. Consequently, thermally activated surface diffusion is more easily suppressed at lower temperatures on (110) surfaces. In addition, surface diffusion is more anisotropic on (110) surfaces than on densely packed (111). In many cases this results in quasi-one-dimensional diffusion, which also hampers self-assembly of two-dimensional islands. Similar to Cu(111), warming up the low-temperature deposited Ag(110) sample to room temperature results in dehalogenation and spontaneous formation of protopolymers.

In summary, only at room temperature are Cu(111) and Ag(110) sufficiently reactive to catalyze homolysis of C–Br bonds in TBB. For low-temperature deposition, differences concerning the mutual arrangement arose: on the densely packed

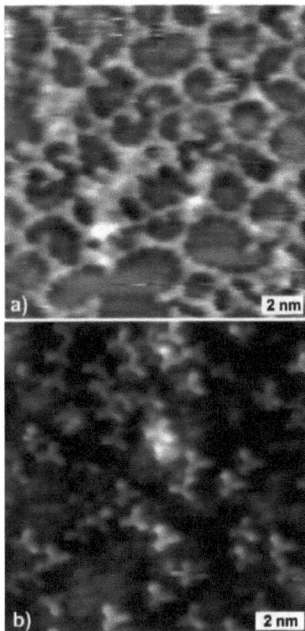

Figure 2. STM topographs of TBB deposited on Ag(110) (a) after warming up the sample to room temperature ($U_T = 1.76$ V, $I_T = 41$ pA) and (b) at ∼80 K ($U_T = -1.50$ V, $I_T = 110$ pA). Low-temperature deposition results in disordered arrangements of single molecules without any indication of ordered self-assembly due to suppressed lateral mobility on the (110) face; warming up the sample to room temperature leads to dehalogenation and the formation of protopolymers similar to the case for Cu(111). Albeit adatoms are not resolved in this case, the measured center-to-center distances of 1.50 nm between interlinked molecules clearly indicates formation of protopolymers.

Cu(111) surface the lateral mobility of TBB is sufficient to facilitate self-assembly into ordered monolayers, while on Ag(110) the lack of surface mobility leads to adsorption of isolated molecules.

Ag(111). In order to gain deeper insight into the relevant parameters for the catalytic activity of coinage metal surfaces for this particular homolysis reaction, further experiments were conducted on Ag(111). Low-temperature deposition onto Ag(111) results in self-assembly of a monolayer structure similar to the Cu(111) case. The arrangement of molecules, the symmetry of the monolayer, and within the experimental error, the lattice parameter are identical for low-temperature deposition on Cu(111) and Ag(111). Most importantly, similar to the aforementioned cases, for low-temperature deposition TBB molecules also stay intact upon adsorption on Ag(111). Yet, qualitatively different observations in comparison to Cu(111) were made when TBB was evaporated onto Ag(111) at room temperature. Instead of protopolymers—a clear indication of dehalogenation—a variety of distinct self-assembled phases based on intact molecules was observed. All structures are stabilized by weak noncovalent interactions between intact molecules. Representative STM topographs of the various phases on Ag(111) are presented in Figure 3.[22] It is noteworthy that low-temperature deposition followed by warming up the sample resulted in the same morphologies as deposition directly at room temperature. In the overview image of Figure 3a, two coexisting phases, namely, a hexagonal and a row structure with oblique unit-cell, can be distinguished. Furthermore, in the lower part of the image molecules that are "frozen" in a disordered state can be identified. We attribute the emergence of this phase to rapid surface diffusion at room temperature, which kinetically traps molecules in the disorder state.

A detailed analysis of the intermolecular distances in the disordered phase does not indicate any formation of protopolymers or covalently interlinked aggregates. Close ups of the ordered phases are presented in Figure 3b,c. Similar to the low-temperature polymorphs observed on Cu(111) and Ag(111), the well-ordered structures are likewise stabilized by electrostatic interactions between nonspherical charge distributions of halogen substituents. Yet, the structures of these polymorphs are more complex, and in addition to the triple halogen—halogen bonds, the hexagonal structure also contains six-membered rings of cyclic halogen bonds, as shown in Figure 3b. Three of those supramolecular hexamers are interconnected via single TBB molecules in a triple Br—Br—Br bond pattern similar to those observed in the low-temperature structure. As illustrated by the overlaid symbolic representations of the molecules, attractive halogen—halogen interactions are topologically very versatile and not restricted to three or six membered rings. Also a slightly displaced head-to-head geometry enables favorable electrostatic interactions, as exemplified in Figure 3c, and gives rise to another polymorph, the row structure. The measured center-to-center distances of halogen—halogen bond associated dimers (2.05 nm) is significantly larger than for metal-coordinated (1.50 nm) or covalently interlinked dimers (1.25 nm)[9] and clearly indicates noncovalent interaction. Figure 4 illustrates the three different types of intermolecular bonding schemes and the corresponding center-to-center distances. For the halogen—halogen interaction, the molecules exhibit a slight lateral displacement and significantly larger center-to-center distance as compared to the covalent and metal coordination case. The substantial differences in center-to-center distance allow for a clear distinction of the interaction type solely based on intermolecular distances as measured in STM topographs.

A displaced halogen—halogen bonded dimer is also the basic unit of the structure depicted in Figure 3d, another occasionally observed polymorph that features a rather high packing density. The experimentally observed coexistence of all structures in Figure 3 indicates the relative weakness and topological versatility of halogen—halogen interactions. In any case, it can be stated that for Ag(111) different noncovalent self-assembled structures were observed dependent on the deposition temperature without any indication of dehalogenation either for room temperature or for low temperature deposition.

In brief, the experiments described above reveal a clear dependency of the catalytic activity on the material [Cu(111) vs Ag(111)], but also on the crystallographic surface orientation [Ag(110) vs Ag(111)].

The experimental findings as summarized in Table 1 give rise to the question about the origin and the detailed mechanism of the catalytic activity of the investigated metal surfaces for the dehalogenation reaction. In particular, we want to address the question of which parameters influence the TBB dehalogenation on coinage metal surfaces. Irrespective of substrate material and orientation, for low-temperature deposition we have never observed dehalogenation, a clear indication of a thermally

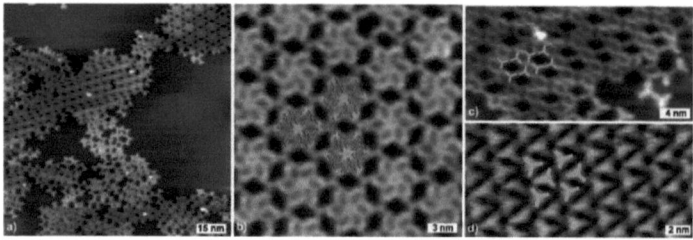

Figure 3. STM topographs of different self-assembled TBB phases on Ag(111). Molecules were deposited at room temperature, while for improved drift stability STM images were acquired at ∼80 K. (a) Overview image presenting two coexisting ordered phases, namely, a row structure on the upper center part and a hexagonal flower structure in the upper right part. The lower half depicts a disordered phase ($U_T = -1.11$ V, $I_T = 102$ pA). (b) Close up of the flower structure with overlaid molecular model ($U_T = -1.11$ V, $I_T = 90$ pA). (c) Close up of the row structure ($U_T = -1.11$ V, $I_T = 112$ pA). (d) Close up of a third, densely packed structure ($U_T = 1.77$ V, $I_T = 94$ pA).

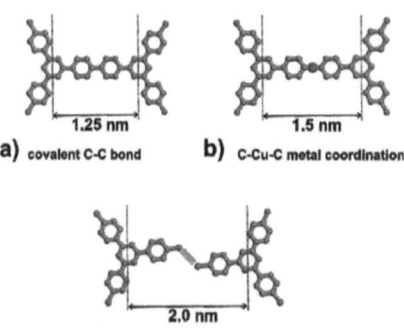

Figure 4. Three possible intermolecular bonding schemes and corresponding center-to-center distances of interlinked molecules. (a) Covalent coupling of TBB molecules leads to the shortest center-to-center distance of ∼1.25 nm. (b) Metal coordination yields a notably higher center-to-center distance of ∼1.50 nm. (c) In addition to cyclic triple halogen–halogen–halogen bonds, also a dimeric arrangement with a center-to-center distance of ∼2.0 nm was observed.

TABLE 1: Summary of Experimentally Observed Intermolecular Bonding Schemes Dependent on Both Substrate Material and Crystallographic Orientation, as Well as Deposition Temperature

	surface @ 80K	surface @ RT
Cu (111)	triple halogen bond	protopolymer
Ag (110)	no ordering	protopolymer
Ag (111)	triple halogen bond	various halogen bond schemes

activated reaction step. TBB molecules do not react and stay intact on the two densely packed Ag(111) and Cu(111) surfaces, where ordered structures were observed. In contrast, on Ag(110) no self-assembly into ordered structures takes place, due to suppressed surface mobility.

In order to understand the substrate dependency of the room-temperature dehalogenation, reactivity is discussed in the framework of heterogeneous catalysis and molecule–metal interactions. Alternatively, an adatom-based surface chemical approach can also explain the experimental findings. Yet, a significant contribution from step edges as active sites for the dehalogenation can be excluded.

Grounded on DFT results, Christensen and Nørskov state that for an accurate description of surface reactivity one has to differentiate between geometrical and electronic effects.[23] In the following, based on their argumentation, we also want to formally distinguish between electronic and geometric effects, where the former can explain the material and the latter the orientation dependency.

The geometrical effect can be explained by means of the active sites concept, which implies that bond cleavage of adsorbates occurs preferentially at low-coordinated surface atoms,[24,25] in particular, at vacancies, kinks, step edges, or dislocations.[8] The literature is rich with examples, where dissociative adsorption favorably occurs at step edges, where reactions rates can be orders of magnitude enhanced as compared to terraces.[26,27]

However, since on Ag(111) dehalogenation has been observed neither for low-temperature nor for room-temperature deposition, it is concluded that the (mostly densely packed) step edges on this surface are not active sites for cleavage of carbon–halogen bonds. Yet, in general, the reactivity of step edges will depend on their crystallographic direction, which determines both the step edge atom coordination and density of kink or ledge sites, giving rise to substantial differences.

Although overview topographs clearly show that many protopolymers are anchored at step edges, we nevertheless exclude a dominant contribution from step edges for the following reason. If the reaction could exclusively proceed at step edges, only step-edge decoration would be observable but not structures extending into terraces. Protopolymers that are bound to step edges would block these active sites and passivate them, resulting in a quenching of the reaction. Such a self-poisoning effect has, for instance, been observed for the dissociation of ethylene on Ni(111) step edges.[28] In conclusion, a dominant contribution from step edges can be ruled out for the dehalogenation reaction.

In the following it is argued that the reactivity differences might originate already in the different atomic arrangement of ideal surfaces and no special active sites are required. While the (111) surfaces of fcc metals are densely packed, the (110) surfaces consist of alternating atomic rows and troughs running along the [$\bar{1}10$] direction. In some respect the (110) surfaces can be seen as a dense stringing of step edges, thereby exposing a large area density of low coordinated surface atoms that promote the catalytic activity. For comparison, the coordination number of a topmost surface atom in an ideal fcc (111) surface is 9 while for an ideal fcc (110) surface the coordination number is only 7. The coordination number can directly affect the energy of the d-band center and thus the reactivity of the respective sites.[29]

A more direct electronic aspect of the catalytical activity comes into play for understanding the observed differences between Cu(111) and Ag(111). Adsorption of aromatic molecules on transition-metal surfaces leads to significant changes in their electronic structure, as concluded for instance from ultraviolet photoelectron spectroscopy (UPS),[30] scanning tunneling spectroscopy (STS),[31] and density functional theory (DFT) studies.[32-34] Depending on the interaction strength, level shifts, level broadening, or emergence of new electronic states due to hybridization are common and most seriously affect frontier molecular orbitals.[35] For instance, Thygesen and Rubio show that the HOMO−LUMO gap of adsorbed molecules shrinks with increasing interaction strength.[36] For the aromatic molecule 3,4,9,10-perylenetetracarboxylic acid dianhydride (PTCDA), it has been shown that fully or partly filled LUMO-derived interface states are created upon adsorption on Cu(111) and Ag(111), respectively, rendering the organic layers semiconducting or metallic. On Au(111), on the other hand, only "soft chemisorption" is reported, where energy levels do not shift significantly because of a relatively weak interaction.[37] These findings are in accordance with the proposed trend of decreasing reactivity for the d10s1 transition metals when moving down this group in the periodic table of the elements from Cu over Ag to Au.[38] This reactivity order was also confirmed by UPS measurements of PTCDA[37] and pentacene[35] on noble metal surfaces. Both compounds serve as model systems for interaction of large π-conjugated molecules with metal surfaces. Since the underlying processes are fundamental and by no means specific for PTCDA or pentacene, it is proposed that this reactivity order can be generalized for other planar π-conjugated aromatic adsorbates. In both cases the newly formed hybridized orbitals originate from interaction of the π-electrons with the metal s- and d-states, in line with the Newns−Anderson model.[30,39] Adsorption of aromatic molecules on transition-metal surfaces is also accompanied by charge transfer between adsorbate and substrate as a consequence of the aforementioned adjusting of the frontier molecular orbitals.[35,37] This can lead to partial filling of mainly the π^* orbital, where the degree of occupancy increases with increasing interaction strength and is thus largest on copper surfaces. Since the π^* orbital is mainly localized at the aromatic system, adsorption induced charge transfer can still not explain the observed homolysis of peripheral C−Br bonds. In order to explain the bond cleavage, we propose that thermally activated charge transfer from the newly occupied π^* into σ^* orbitals, which are antibonding with respect to the C−Br bond, eventually destabilizes these bonds and facilitates homolysis. A similar two-step mechanism for C−X bond dissociation in solution was found by Kimura and Takamuku, who studied halogen scission in aryl halides[40] and benzyl halides[41] by means of low-temperature pulse radiolysis. First, an additional electron is captured by the π^* orbital and then in a second step transferred into the C−halogen σ^* orbital. This results in destabilization and dissociation of the C−Br bond. Moreover, a comparable two-step mechanism is also discussed for photodissociation of dibromobenzene and tribromobenzene, where the initially excited singlet (π, π^*) state in the phenyl ring decays into the repulsive triplet (n,σ^*) state located at the C−Br bond.[42] On the basis of these findings we conclude that the dehalogenation of TBB on densely packed noble metal surfaces can only occur when the interaction strength is sufficiently strong. Evidently, this criterion is fulfilled for Cu(111), but not for Ag(111). For higher corrugated (110) surfaces, however, the reactivity of Ag(110) becomes sufficient to catalyze the dehalogenation reaction. Along the lines of heterogeneous catalysis research, the higher reactivity of Ag(110) as compared to Ag(111) is explained with the higher surface corrugation leading to a lower coordination and thus higher reactivity of surface atoms. Again, Zou and co-workers have confirmed this trend for the adsorption of PTCDA on Ag(111) and Ag(110), where the more corrugated (110) face exhibits stronger interaction.[30]

As already stated above, a decisive influence of adatom chemistry would also be consistent with our experimental observations and cannot be fully excluded. It is well-known that for metal surface chemistry adatoms can be important mediators or reaction partners for various types of reactions.[29] Consequently, both the temperature and surface dependent density of the adatom gas as well as the adatom reactivity can explain reactivity differences. For instance, a face-specific dependency for the adsorption geometry of benzoate molecules has been attributed to the availability of metal adatoms, being significantly higher on Cu(110) as compared to Cu(111).[43] It has also been reported that the deprotonation of carboxylic groups in trimesic acid molecules does not take place on pristine Ag(111) at room temperature but can be triggered by an additional supply of more reactive copper atoms.[19] Especially the latter example exemplifies the importance of adatom chemistry for the formation of metal-coordination complexes on surfaces. In this picture, the temperature dependence can be explained by suppression of the adatom gas at lower temperature, while the orientation dependence can be explained by different binding energies of atoms in step edges. However, since we do not observe any formation of protopolymers on Ag(111), a dominant contribution from adatoms for the dehalogenation reaction seems unlikely. Even though the density of adatoms on fcc(111) surfaces is substantially lower than on (110) surfaces, as rationalized by a model based on the change of coordination number for the detachment process,[43] at least a few coordination complexes should also be observable on Ag(111), if the reaction was exclusively driven by adatoms. However, a plain consideration of merely the adatom density is not satisfying, and also the adatom reactivity has to be considered. Since the coordination number of adatoms is also surface-dependent, it is conceivable that Ag adatoms behave chemically distinctly on (111) than on (110) surfaces. In order to obtain a detailed and quantitative understanding of the dehalogenation reaction, theoretical studies that address the electronic structure of the chemisorbed molecule−substrate complex and tackle conceivable reactivity differences of adatoms are very desirable.

Conclusions

In summary, studies of a heterogeneously catalyzed dehalogenation reaction, namely full debromination of the aromatic compound TBB, on single crystal metal surfaces revealed interesting reactivity differences. Since the reaction only proceeds on Cu(111), but not on Ag(111), the catalytic capability of the substrate for this reaction is clearly material dependent. On the other hand, the dehalogenation reaction took place on Ag(110), thereby exemplifying that also the specific surface orientation can be decisive. Third, in variable-temperature experiments it was found that the dehalogenation reaction cannot proceed at low substrate temperatures (~80 K) irrespective of the substrate, thereby proving the necessity of thermal activation. In order to explain the occurrence of the reaction as a function of different experimental parameters, we propose a two-step mechanism, where initial charge transfer upon adsorption provides the basis for occupation of an antibonding orbital. Besides the recognized role of active sites, this comparative series of experiments elucidates that the overall reactivity of a catalytically active surface originates from a combination of atomic arrangement and electronic structure.

For future experiments, it would also be enlightening to study the role of the organic compound and its respective electronic structure. For instance, the HOMO−LUMO gap can be altered by means of decreasing or increasing the size of the aromatic system, thereby also affecting the level alignment and magnitude of charge transfer. The strength of the carbon−halogen bond is another accessible parameter worthy of study. This bond can be weakened by substituting bromine with iodine, but it can also be strengthened by substituting bromine with chlorine.

Methods

All samples were prepared and characterized in an ultrahigh vacuum chamber (base pressure <5 × 10^{-10} mbar) equipped with a scanning tunneling microscope (STM). Metal single crystals [Cu(111), Ag(111), and Ag(110)] were prepared by repeated cycles of Ne$^+$ ion sputtering and annealing. 1,3,5-Tris(4-bromophenyl)benzene was obtained from a commercial source (Sigma Aldrich) and vacuum sublimed from a home-built Knudsen cell with crucible temperatures between 150 and 160 °C.[44] Samples were deposited in the microscope, which is a home-built beetle-type STM mounted on a flow cryostat and thus able to operate at variable temperatures. During deposition the STM and the substrates were held either at room temperature or at ∼80 K. Typically, images were acquired at ∼80 K (also for room temperature deposition), because of improved drift stability of the instrument at low temperatures.

Acknowledgment. Financial support by the Deutsche Forschungsgemeinschaft (SFB 486) and the Nanosystems Initiative Munich (NIM) is gratefully acknowledged. G.E. acknowledges support by the Hanns-Seidel-Stiftung.

References and Notes

(1) Ertl, G.; Freund, H. J. *Phys. Today* **1999**, *52*, 32−38.
(2) Ertl, G. *J. Vac. Sci. Technol., A* **1983**, *1*, 1247−1253.
(3) Bowker, M. *Surf. Sci.* **2009**, *603*, 2359−2362.
(4) Bowker, M. *Chem. Soc. Rev.* **2007**, *36*, 1656−1673.
(5) Vang, R. T.; Lauritsen, J. V.; Laegsgaard, E.; Besenbacher, F. *Chem. Soc. Rev.* **2008**, *37*, 2191−2203.
(6) Leibsle, F. M.; Murray, P. W.; Francis, S. M.; Thornton, G.; Bowker, M. *Nature* **1993**, *363*, 706−709.
(7) Africh, C.; Comelli, G. *J. Phys.: Condens. Matter* **2006**, *18*, R387−R416.
(8) Zambelli, T.; Wintterlin, J.; Trost, J.; Ertl, G. *Science* **1996**, *273*, 1688−1690.
(9) Gutzler, R.; Walch, H.; Eder, G.; Kloft, S.; Heckl, W. M.; Lackinger, M. *Chem. Commun.* **2009**, 4456−4458.
(10) Sykes, E. C. H.; Han, P.; Kandel, S. A.; Kelly, K. F.; McCarty, G. S.; Weiss, P. S. *Acc. Chem. Res.* **2003**, *36*, 945−953.
(11) McCarty, G. S.; Weiss, P. S. *J. Am. Chem. Soc.* **2004**, *126*, 16772−16776.
(12) Lipton-Duffin, J. A.; Ivasenko, O.; Perepichka, D. F.; Rosei, F. *Small* **2009**, *5*, 592−597.
(13) Xi, M.; Bent, B. E. *J. Am. Chem. Soc.* **1993**, *115*, 7426−7433.
(14) Ullmann, F.; Bielecki, J. *Ber. Dtsch. Chem. Ges.* **1901**, *34*, 2174−2185.
(15) Migani, A.; Illas, F. *J. Phys. Chem. B* **2006**, *110*, 11894−906.
(16) Nanayakkara, S. U.; Sykes, E. C. H.; Fernandez-Torres, L. C.; Blake, M. M.; Weiss, P. S. *Phys. Rev. Lett.* **2007**, *98*, 206108.
(17) Blake, M. M.; Nanayakkara, S. U.; Claridge, S. A.; Fernandez-Torres, L. C.; Sykes, E. C. H.; Weiss, P. S. *J. Phys. Chem. A* **2009**, *113*, 13167−13172.
(18) Pai, W. W.; Bartelt, N. C.; Peng, M. R.; Reuttrobey, J. E. *Surf. Sci.* **1995**, *330*, L679−L685.
(19) Lin, N.; Payer, D.; Dmitriev, A.; Strunskus, T.; Woll, C.; Barth, J. V.; Kern, K. *Angew. Chem., Int. Ed.* **2005**, *44*, 1488−1491.
(20) Bosch, E.; Barnes, C. L. *Cryst. Growth Des.* **2002**, *2*, 299−302.
(21) Awwadi, F. F.; Willett, R. D.; Haddad, S. F.; Twamley, B. *Cryst. Growth Des.* **2006**, *6*, 1833−1838.
(22) In order to improve the drift stability of the microscope, measurements were performed at ∼80 K, also for samples where molecules were deposited at room temperature.
(23) Christensen, C. H.; Norskov, J. K. *J. Chem. Phys.* **2008**, *128*, 182503.
(24) Taylor, H. S. *Proc. R. Soc. London, Ser. A* **1925**, *108*, 105−111.
(25) Norskov, J. K.; Bligaard, T.; Hvolbaek, B.; Abild-Pedersen, F.; Chorkendorff, I.; Christensen, C. H. *Chem. Soc. Rev.* **2008**, *37*, 2163−2171.
(26) Gambardella, P.; Sljivancanin, Z.; Hammer, B.; Blanc, M.; Kuhnke, K.; Kern, K. *Phys. Rev. Lett.* **2001**, *87*, 056103.
(27) Liu, Z. P.; Hu, P. *J. Am. Chem. Soc.* **2003**, *125*, 1958−1967.
(28) Vang, R. T.; Honkala, K.; Dahl, S.; Vestergaard, E. K.; Schnadt, J.; Laegsgaard, E.; Clausen, B. S.; Norskov, J. K.; Besenbacher, F. *Nat. Mater.* **2005**, *4*, 160−162.
(29) Hammer, B. *Top Catal* **2006**, *37*, 3−16.
(30) Zou, Y.; Kilian, L.; Scholl, A.; Schmidt, T.; Fink, R.; Umbach, E. *Surf. Sci.* **2006**, *600*, 1240−1251.
(31) Kilian, L.; Hauschild, A.; Temirov, R.; Soubatch, S.; Scholl, A.; Bendounan, A.; Reinert, F.; Lee, T. L.; Tautz, F. S.; Sokolowski, M.; Umbach, E. *Phys. Rev. Lett.* **2008**, *100*, 136103.
(32) Bilic, A.; Reimers, J. R.; Hush, N. S.; Hoft, R. C.; Ford, M. J. *J. Chem. Theory Comput.* **2006**, *2*, 1093−1105.
(33) Toyoda, K.; Nakano, Y.; Hamada, I.; Lee, K.; Yanagisawa, S.; Morikawa, Y. *J. Electron Spectrosc. Relat. Phenom.* **2009**, *174*, 78−84.
(34) Romaner, L.; Nabok, D.; Puschnig, P.; Zojer, E.; Ambrosch-Draxl, C. *New J. Phys.* **2009**, *11*, 053010.
(35) Yamane, H.; Kanai, K.; Ouchi, Y.; Ueno, N.; Seki, K. *J. Electron Spectrosc. Relat. Phenom.* **2009**, *174*, 28−34.
(36) Thygesen, K. S.; Rubio, A. *Phys. Rev. Lett.* **2009**, *102*, 046802.
(37) Duhm, S.; Gerlach, A.; Salzmann, I.; Broker, B.; Johnson, R. L.; Schreiber, F.; Koch, N. *Org. Electron.* **2008**, *9*, 111−118.
(38) Hammer, B.; Norskov, J. K. *Nature* **1995**, *376*, 238−240.
(39) Tautz, F. S. *Prog. Surf. Sci.* **2007**, *82*, 479−520.
(40) Kimura, N.; Takamuku, S. *J. Am. Chem. Soc.* **1995**, *117*, 8023−8024.
(41) Kimura, N.; Takamuku, S. *Bull. Chem. Soc. Jpn.* **1993**, *66*, 3613−3617.
(42) Kadi, M.; Davidsson, J. *Chem. Phys. Lett.* **2003**, *378*, 172−177.
(43) Perry, C. C.; Haq, S.; Frederick, B. G.; Richardson, N. V. *Surf. Sci.* **1998**, *409*, 512−520.
(44) Gutzler, R.; Heckl, W. M.; Lackinger, M. *Rev. Sci. Instrum.* **2010**, *81*, 015108.

JP102704Q

7.3 A combined ion-sputtering and electron-beam annealing device for the *in-vacuo* post preparation of scanning probes

Georg Eder, Stefan Schlögl, Klaus Macknapp, Wolfgang M. Heckl, and Markus Lackinger

Rev. Sci. Instrum., 82, 033701-01 (**2011**),
selected for March 14, 2011 issue of Virtual Journal of
Nanoscale Science & Technology
http://dx.doi.org/10.1063/1.3556443

Reprinted with permission from[103]. Copyright 2011, American Institute of Physics

A combined ion-sputtering and electron-beam annealing device for the *in vacuo* postpreparation of scanning probes

Georg Eder,[1] Stefan Schlögl,[1] Klaus Macknapp,[2] Wolfgang M. Heckl,[2,3] and Markus Lackinger[1,2]

[1]*Department of Earth and Environmental Sciences and Center for NanoScience (CeNS), Ludwig-Maximilians-University, Theresienstrasse 41, 80333 Munich, Germany*
[2]*Deutsches Museum, Museumsinsel 1, 80538 Munich, Germany*
[3]*Department of Physics, TUM School of Education, Technical University Munich, Schellingstrasse 33, 80333 Munich, Germany*

(Received 29 November 2010; accepted 29 January 2011; published online 1 March 2011)

We describe the setup, characteristics, and application of an *in vacuo* ion-sputtering and electron-beam annealing device for the postpreparation of scanning probes (e.g., scanning tunneling microscopy (STM) tips) under ultrahigh vacuum (UHV) conditions. The proposed device facilitates the straightforward implementation of a common two-step cleaning procedure, where the first step consists of ion-sputtering, while the second step heals out sputtering-induced defects by thermal annealing. In contrast to the standard way, no dedicated external ion-sputtering gun is required with the proposed device. The performance of the described device is demonstrated by SEM micrographs and energy dispersive x-ray characterization of electrochemically etched tungsten tips prior and after postprocessing. © *2011 American Institute of Physics*. [doi:10.1063/1.3556443]

I. INTRODUCTION

For scanning probe microscopy experiments the scanning probe, e.g., the metal tip for STM, is of pivotal importance. The tip is crucial for the stability of the tunneling current, and can thus be decisive for the noise levels of both signal and data. Yet, since the STM imaging process convolutes geometric and electronic properties of sample and tip, the probe also has a great impact on the contrast in high-resolution topographs, and similarly on spectroscopic data. Although the atomic configuration at the apex can normally not be controlled, at least the chemical cleanliness of the tips should be guaranteed. For the majority of STM experiments under UHV conditions tungsten is still the material of choice for the tip, not at least because electrochemical etching techniques for sharp tips are well developed.[1,2] The mechanical properties of tungsten and its high melting point definitely render the material suitable; however, electrochemically etched tungsten tips are prone to oxidation under ambient conditions.[1,2] Oxidized tungsten tips can result in unstable tunneling conditions and poor image quality, or the insulating oxide coating can even cause tip crashes during coarse approach.[3] In order to remove not only tungsten-oxide layers but also etching remnants and byproducts, various procedures are proposed in the literature like annealing by electron bombardment,[4,5] ion-sputtering,[6–8] dipping into hydrofluoric acid (HF),[9] and self-sputtering in a noble gas environment.[10] Although it is difficult to establish a clear correlation between a specific postpreparation procedure and achieved quality of STM data, there is a consensus that an appropriate after-treatment improves the overall performance of STM tips.[3] For STM experiments under UHV conditions the standard tip after-treatment is *in situ* ion-sputtering and subsequent thermal annealing. For sample preparation most UHV systems are equipped with an ion-sputtering gun. Due to the operation principle of STM the sample surface and the tip point into opposite directions. Hence, either a manipulation mechanism to rotate the tip or a second (typically rather expensive) ion-sputtering gun is necessary. Here, as an alternative we present a versatile, inexpensive, easy to realize, and customizable setup for the *in vacuo* postpreparation of scanning probes. First the setup is described, and then we present a characterization of the device by measuring the ion current at the tip as a function of various parameters along with a finite element simulation of the electrostatic potential and field. Finally, we demonstrate the efficiency of the device by comparison of as-etched with postprocessed tungsten tips.

II. EXPERIMENTAL SETUP

The basic setup of the postpreparation device is depicted in Fig. 1(a) and consists of an axial arrangement of filament, ring, and tip. In this test setup the tip is held by a spring-loaded socket (accepting wire diameters from 0.35 to 0.55 mm). The UHV compatible setup is assembled on a 40 CF flange equipped with four SHV high-voltage feedthroughs (rated for up to 5 kV and 16.5 A). The filament is made from a tungsten wire (diameter 0.2 mm, 13 coils), where one end is grounded to the flange on the vacuum side, while the other end is connected to one of the feedthroughs. The ring (tungsten wire, wire diameter 0.2 mm, ring diameter ~10 mm) is not closed and both ends are likewise connected to SHV feedthroughs. The fourth feedthrough is connected to the tip (holder). The heights of ring and filament on the axis are adjustable and distances are set to 12 mm between tip and ring, and 16 mm between ring and filament for the experiments described. The proposed basic setup can easily be customized to a specific tip transfer and carrier system. Figure 1(b) presents a sketch of a conceivable (not realized) adaptation of the

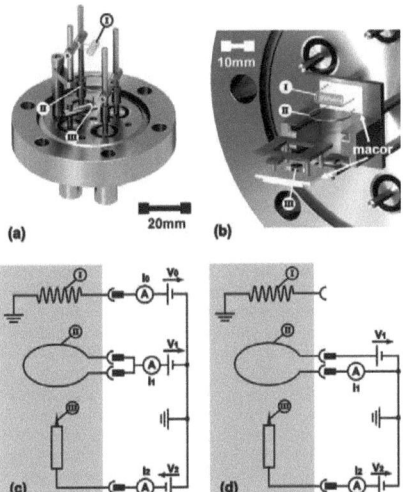

FIG. 1. (Color online) (a) Scheme of the combined ion-sputtering and electron-beam annealing device: (I) filament, (II) ring used as grid for ion-sputtering and used as a filament for electron-beam annealing, and (III) tip holder. (b) Adaptation scheme of the proposed device for Omicron tip holders. (c) Wiring diagram for the ion-sputtering mode: V_0 = filament voltage, V_1 = acceleration voltage for electrons, V_2 = acceleration voltage for noble gas ions, I_0 = filament current, I_1 = electron emission current, I_2 = ion current (d) Wiring diagram for the electron-beam annealing mode: V_1 = filament voltage, V_2 = acceleration voltage for electrons, I_1 = filament current, I_2 = electron emission current.

device for standard Omicron tip holders. The base plate of this slightly adapted Omicron tip holder is now made from macor —a machinable sinter ceramic. Also, the standard glued-on magnet is replaced by a mechanically clamped magnet with high Curie temperature, in order to prevent it from falling off or demagnetization during tip annealing. The tip is electrically contacted from below with a sliding contact. The upper plate of the Omicron tip holder exhibits a wide cutout to avoid charging effects. The dimension and material of filament and ring are identical to the actually tested setup in Fig. 1(a). Ring and filament assemblies are fixed by setscrews in an insulating macor block and contacted from behind to electrical feedthroughs. This macor block is mounted onto a stainless steel fixture. The tip preparation stage can simply be loaded and unloaded by manipulation of Omicron tip holders into mountings on the side. Comparable adaptations are conceivable to any systems which feature tip exchange.

The proposed device can be operated in two different modes: ion-sputtering and electron-beam annealing. In order to alternate between these two modes, only the external wiring has to be changed, and respective wiring diagrams are depicted in Figs. 1(c) and 1(d).

For the ion-sputtering mode, a noble gas (typically neon or argon) has to be introduced into the vacuum chamber as the sputtering gas. Since most UHV systems are equipped with an ion-sputtering gun for sample preparation, the associated variable precision leak-valve can be used and no additional installations are required. After a partial noble gas pressure in the order of 10^{-5} mbar has been established, a dc current (~4.5 A) is passed through the filament, yielding thermal emission of electrons (~10 mA). While the filament is grounded, a positive voltage in the order of +0.8 kV is applied to the ring. Thermally emitted electrons are accelerated toward the ring and generate positive noble gas ions through impact ionization. In order to accelerate the positive ions toward the scanning probe, a negative voltage on the order of −2.0 kV is applied to the tip, resulting in a measurable ion current in the magnitude of microamperes. The ion current increases with both noble gas pressure and negative voltage on the tip as depicted in Fig. 2.

For ion-sputtering of tips with an external ion-sputtering gun best results are obtained when tips are mounted at the center of the ion beam.[7] The required spherical symmetry is inherently achieved by design in our setup. Generally ion milling processes are dependent on many parameters such as the ion's angle of incidence,[6] their kinetic energy, and their

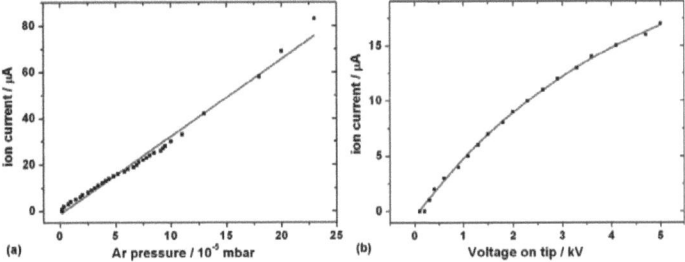

FIG. 2. Characteristics of the proposed device: (a) Ion current I_2 as a function of partial Ar pressure (I_0 = 10 mA (emission), V_2 = −2.0 kV (tip), V_1 (ring) was adjusted to keep I_0 (emission) constant). In the pressure range from 1×10^{-5} to 2.5×10^{-4} mbar the ion current as measured at the tip increases approximately linearly with Ar pressure, (b) ion current as a function of acceleration voltage at the tip ($p_{Ar} = 4 \times 10^{-5}$ mbar, I_0 = 10 mA (emission), V_1 (ring) adjusted to keep I_0 (emission) constant). An increasing negative extraction voltage at the tip results in sublinear increase of the ion current. Solid lines serve as guides to the eye.

chemical nature. In the proposed setup, the kinetic energy can easily be adjusted, while the angle of incidence apparently depends on the local surface orientation.

The second mode of the device is electron-beam annealing of the tip. For this purpose the ring which was used as a grid in the ion-sputtering mode is now used as filament, i.e., as a thermal electron source. As can be seen in the wiring diagram in Fig. 1(d) one side of the ring is grounded at the atmospheric side now, and a positive voltage on the order of +1.5 kV is applied to the tip. The ring when operated as a filament is heated by a dc current of ∼4 A, resulting in an emission current of 1.5 mA as measured at the biased tip. The upper original filament is too remote to yield a reasonable electron current when a positive voltage is applied to the tip, possibly because of electrostatic screening through the ring. Annealing of sputtered STM tips is necessary to heal out sputtering-induced defects, yet, the annealing power as the product of emission current and applied voltage to the tip must not be too high to prevent melting-induced blunting of the apex.[4] The enhancement of the electrostatic field due to the low radius of curvature at the tip leads to effective local heating of its apex.

For a quantitative understanding of the proposed device where a high positive voltage is applied to the ring and in close vicinity a high negative voltage is applied to the tip, a finite-element simulation of the electrostatic potential and electric field distribution were performed with the program package Ansoft's MAXWELL 2D.[11] Since an idealized setup exhibits rotational symmetry, resulting potential and field distributions are likewise rotational symmetric, and the 2D solution in the median plane represents a cross section of the 3D solution in the r–z plane. The geometry has been taken from the experiment, and simulations results for the ion-sputtering mode with −2.0 kV at the tip, +0.8 kV at the ring, and the filament grounded are depicted in Fig. 3. These simulations confirm

that the electric field vectors point toward the tip apex; hence, ions are accelerated to this region. Also the positively biased ring effectively screens the negative potential from the tip, and is thus able to attract electrons.

III. CHARACTERIZATION OF UNTREATED AND POSTPROCESSED STM TIPS

In order to demonstrate the efficiency of the proposed device, electrochemically etched tungsten tips were characterized by scanning electron microscopy (SEM, Zeiss LEO 440i) and spatially averaged energy dispersive x-ray (EDX) analysis. Topographs and spectra of the same tips were acquired directly after etching and compared to measurements acquired after ion-sputtering (Fig. 4) and annealing (Fig. 5). The STM tips were initially prepared by electrochemical ac etching of a polycrystalline tungsten wire (diameter 0.5 mm) in aqueous 2M KOH solution. In a second step, these tips were sharpened by electropolishing under optical control in a light microscope.[12] Ion-sputtering was performed with Ar$^+$ ions for 1 and 5 min, respectively, with the following parameters: ∼1.5 × 10^{-5} mbar (3.1 × 10^{-5} mbar) Ar pressure, 4.0 A filament current, 10 mA electron emission current, +830 V ring voltage, −2.0 kV tip voltage. These values result in a stable ion current of 5 μA (10 μA) at the tip. Directly after electropolishing many tips exhibit not further identified, but clearly visible, contaminations [cf. Figs. 4(a) and 4(c)]. For about 50% of the tips, oxygen was detected by EDX at the foremost part, and attributed to the presence of tungsten oxide. These EDX-supported findings are in accordance with transmission electron microscopy studies by Garnaes et al.,[13] who also concluded that electrochemically etched tungsten tips are covered with a few nanometer thick oxide layers. Similarly by means of EDX, aluminum was detected at the shank of the tip where the tungsten has not been

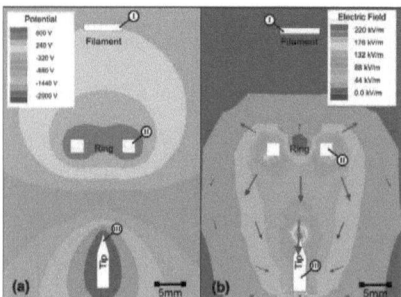

FIG. 3. (Color online) Finite element simulation results of (a) electrostatic potential and (b) electrostatic field (vectors and magnitudes) distributions for a 2D cross-sectional geometry of the decive in the ion-sputtering mode with filament grounded, +0.8 kV at the ring, and −2.0 kV at the tip. These 2D maps represent a cross section of the rotational symmetric 3D solution in the r–z plane. Positively charged noble gas ions are generated above the ring by electron impact ionization, and are accelerated toward the tip by the electric field below the ring.

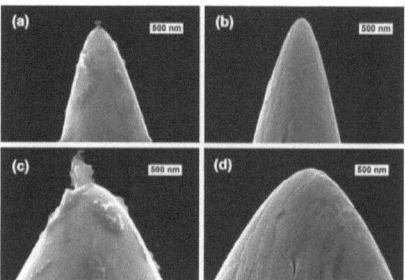

FIG. 4. SEM micrographs of electrochemically etched tungsten tips (a)/(c) directly after electrochemical etching and electropolishing without any further treatment, (b) the same tip as shown in (a) after sputtering with 5 μA for 1 min, (d) the same tip as shown in (c) after sputtering with 10 μA for 5 min. Both examples clearly demonstrate that ion-sputtering in the proposed device is efficient in removing contaminations, but also changes the surface structure, and possibly the tip shape. The tip shown in (d) was sputtered with a tenfold increased ion-dose as compared to (b), which already resulted in a detectable change of outer shape.

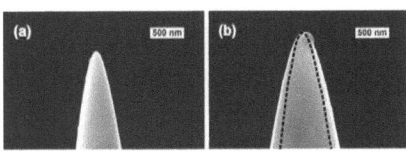

FIG. 5. SEM micrographs of an electrochemically etched tungsten tip (a) before and (b) after electron beam annealing (+1.5 kV, 1.5 mA, i.e., 2.25 W for 300 s). The outer shape of the tip before annealing has been reproduced in (b) by the dashed line, in order to illustrate the melting induced change of shape. The cone angle changes from 20° to 25°.

electrochemically etched. These contaminations might originate from the wire drawing process. After the postpreparation through ion-sputtering each tip was characterized again by SEM and EDX [cf. Figs. 4(b) and 4(d) for representative examples]. Apparently all contaminations have been removed and no oxide was detected anymore in the EDX spectra of ion-sputtered tips. As is evident from Fig. 4(d) the topography changes and the sputtered surfaces become rougher. Sputtering-induced surface roughening can be reduced by lowering the kinetic energy of the Ar$^+$ ions to 1 keV or less.[7]

Comparison of SEM images obtained for different sputtering times reveals this postpreparation method as a time critical process. The longer the sputtering time, the more material is removed. In order to smoothen the surface and heal out defects after the ion-sputtering treatment, cycles of heating are recommended.

Tip annealing was also found to be a time critical process, and moreover critical parameters for blunting are highly dependent on the initial microscopic shape of the tip, which is uncontrollably influenced not only by the electrochemical fabrication process but also by ion-sputtering. Typical parameters for tip annealing are high voltages around +1 kV, emission currents of ∼1 mA, and a time span of minutes, where 1 min is more on the conservative side. SEM micrographs of an electropolished STM tip before and after annealing (+1.5 kV, 1.5 mA, 5 min) are depicted in Figs. 5(a) and 5(b), respectively. Comparison of the outer shapes reveals that this particular tip was already partly melted by the annealing process and became blunted. Even if the electrochemical etching procedure is carried out in a similar way with similar parameters, there is nevertheless a large scatter in the cone angle of the tips. Thus, it is difficult if not impossible to provide general parameters for annealing of tips. Nevertheless, we propose that for cone angles ∼25°, no indications of blunting were observable in SEM micrographs for annealing times <3 min and currents of 1.5 mA at high voltages of +1.5 kV.

IV. CONCLUSION

We presented a device which can be used for both ion-sputtering and electron-beam annealing of STM tips without the need to change the hardware or manipulate the tip between both modes. Its performance is demonstrated by SEM images of sputtered tips which are free from previously detected contaminations, but exhibit clearly increased surface roughness. In addition, oxygen is absent in the EDX spectra of postprocessed tips, pointing toward complete removal of the tungsten oxide layer by the sputtering treatment.

Although self-sputtering (i.e., a high negative voltage applied to a sharp tip causes field emission; in a gas atmosphere, field emitted electrons can generate positive ions which are then accelerated toward the tip and sputter) is also a straightforward method to sputter STM tips, it imposes requirements on the tip. Only tips which are already sharp enough for field emission are suitable for self-sputtering. Since the proposed device also works for blunt tips it might be particularly useful to recover STM tips in UHV systems without a load lock and the possibility to introduce new tips.

ACKNOWLEDGMENTS

Financial support by the Nanosystems-Initiative Munich (NIM) is gratefully acknowledged. S.S. acknowledges financial support by Elitenetzwerk Bayern. G.E. is particularly grateful for financial support by the Hanns-Seidel-Stiftung. We would also like to thank Stephan Kloft for his help with the drawings.

[1] J. P. Ibe, P. P. Bey, S. L. Brandow, R. A. Brizzolara, N. A. Burnham, D. P. DiLella, K. P. Lee, C. R. K. Marrian, and R. J. Colton, J. Vac. Sci. Technol. A **8**, 3570 (1990).
[2] A. D. Müller, F. Müller, M. Hietschold, F. Demming, J. Jersch, and K. Dickmann, Rev. Sci. Instrum. **70**, 3970 (1999).
[3] I. Ekvall, E. Wahlstrom, D. Claesson, H. Olin, and E. Olsson, Meas. Sci. Technol. **10**, 11 (1999).
[4] N. Ishida, A. Subagyo, A. Ikeuchi, and K. Sueoka, Rev. Sci. Instrum. **80**, 093703 (2009).
[5] Z. Q. Yu, C. M. Wang, Y. Du, S. Thevuthasan, and I. Lyubinetsky, Ultramicroscopy **108**, 873 (2008).
[6] P. Hoffrogge, H. Kopf, and R. Reichelt, J. Appl. Phys. **90**(10), 5322 (2001).
[7] S. Morishita and F. Okuyama, J. Vac. Sci. Technol. A **9**, 167 (1991).
[8] R. Zhang and D. G. Ivey, J. Vac. Sci. Technol. B **14**, 1 (1996).
[9] E. Paparazzo, L. Moretto, S. Selci, M. Righini, and I. Farne, Vacuum **52**, 421 (1999).
[10] S. Ernst, S. Wirth, M. Rams, V. Dolocan, and F. Steglich, Sci. Technol. Adv. Mat. **8**, 347 (2007).
[11] Ansoft MAXWELL 2D Version 9.0.573SV, Available: www.ansoft.com.
[12] J. T. Yates, *Experimental Innovations in Surface Science: A Guide to Practical Laboratory Methods and Instruments* (Springer-Verlag, New York, 1998).
[13] J. Garnaes, F. Kragh, K. A. Morch, and A. R. Tholen, J. Vac. Sci. Technol. A **8**(1), 441 (1990).

PUBLICATION

7.4 Extended two-dimensional metal-organic frameworks based on thiolate-copper coordination bonds

Hermann Walch, Jürgen Dienstmaier, Georg Eder, Rico Gutzler,
Stefan Schlögl, Thomas Sirtl, Kalpataru Das,
Michael Schmittel, and Markus Lackinger
J. Am. Chem. Soc., **2011**, 133, 7909–7915
http://dx.doi.org/10.1021/ja200661s

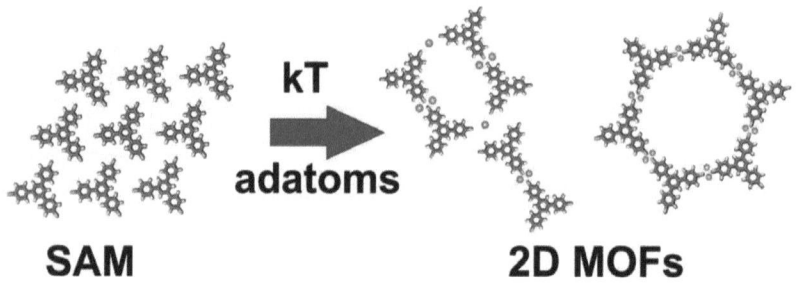

Reprinted with permission from [238]. Copyright 2011 American Chemical Society

Extended Two-Dimensional Metal−Organic Frameworks Based on Thiolate−Copper Coordination Bonds

Hermann Walch,[†] Jürgen Dienstmaier,[†] Georg Eder,[†] Rico Gutzler,[†,#] Stefan Schlögl,[†] Thomas Sirtl,[⊥] Kalpataru Das,[‡] Michael Schmittel,[‡] and Markus Lackinger*[,†,§,⊥]

[†]Department for Earth and Environmental Sciences and Center for NanoScience, Ludwig-Maximilians-Universität, Theresienstrasse 41, 80333 München, Germany

[‡]Center of Micro and Nanochemistry and Engineering, Organische Chemie I, Universität Siegen, Adolf-Reichwein-Strasse 2, 57068 Siegen, Germany

[§]Deutsches Museum, Museumsinsel 1, 80538 München, Germany

[⊥]Technical University Munich, TUM School of Education, Schellingstrasse 33, 80799 München, Germany

ABSTRACT: Self-assembly and surface-mediated reactions of 1,3,5-tris(4-mercaptophenyl)benzene—a three-fold symmetric aromatic trithiol—are studied on Cu(111) by means of scanning tunneling microscopy (STM) under ultrahigh-vacuum (UHV) conditions. In order to reveal the nature of intermolecular bonds and to understand the specific role of the substrate for their formation, these studies were extended to Ag(111). Room-temperature deposition onto either substrate yields densely packed trigonal structures with similar appearance and lattice parameters. Yet, thermal annealing reveals distinct differences between both substrates: on Cu(111) moderate annealing temperatures (∼150 °C) already drive the emergence of two different porous networks, whereas on Ag(111) higher annealing temperatures (up to ∼300 °C) were required to induce structural changes. In the latter case only disordered structures with characteristic dimers were observed. These differences are rationalized by the contribution of the adatom gas on Cu(111) to the formation of metal-coordination bonds. Density functional theory (DFT) methods were applied to identify intermolecular bonds in both cases by means of their bond distances and geometries.

■ INTRODUCTION

In the past years a great structural and chemical variety of surface-supported metal−organic networks has been demonstrated by combining appropriately functionalized organic building blocks with various coordinating metals.[1] While some of the coordination complexes utilized for surface-confined systems were already well-known from bulk systems, other coordination numbers and geometries are unique to surface-supported networks. Concerning the intermolecular bond strength, and thus the overall stability of the structures, metal−organic networks occupy an intermediate position between hydrogen-bonded networks and covalent organic frameworks.[1a] Yet, since metal-coordination bonds are reversible under commonly applied growth conditions, the preparation of long-range ordered networks becomes feasible. The motivation of this work is to extend the tool box of functional groups for the design of surface-supported metal−organic networks to thiol groups and understand the formation kinetics and topological properties of thiolate−metal complexes. To this end we designed and synthesized a highly symmetric aromatic trithiol molecule and studied its self-assembly and surface-supported reactions on a Cu(111) surface, which is known to inherently supply copper coordination centers from its free adatom gas.[2] To clarify the specific role of the substrate, similar experiments were also conducted on Ag(111). Thiolate−copper coordination bonds are of particular interest because of their electronic conjugation which allows electronically coupling of molecular units by thiolate−copper−thiolate bonds.[3] One conceivable application of copper−thiolate complexes hence lies in the field of molecular electronics, where reliable tools are required for interconnection of molecular entities in an atomically defined manner without perturbing or interrupting electronic conjugation.

■ EXPERIMENTAL DETAILS

All samples were prepared and characterized in an ultrahigh-vacuum chamber (base pressure <5 × 10^{-10} mbar) equipped with an Omicron VT scanning tunneling microscope (STM). Cu(111) single crystals were prepared by subsequent cycles of Ar$^+$-ion sputtering and annealing at 820 K. Additional low energy electron diffraction (LEED) experiments for further characterization of the precursor structure were carried out in a separate UHV system equipped with standard preparation facilities and LEED optics from Omicron. LEED measurements were

Received: February 8, 2011
Published: May 02, 2011

Figure 1. (a) TMB fully deprotonates upon room temperature adsorption on reactive copper surfaces forming a surface-anchored trithiolate. (b) STM topograph of as-deposited TMB on Cu(111) acquired at room temperature (I_T = 185 pA, U_T = 0.79 V, 10 × 10 nm², $a = b = 1.30$ nm, $\gamma = 120°$, unit cell indicated by dashed white lines). The densely packed trigonal structure contains one molecule per unit cell. (c) Tentative model of the densely packed trithiolate structure including the Cu(111) substrate. While the azimuthal orientation of the TMB-derived trithiolates with respect to the substrate directions can be inferred from the experiment, the precise adsorption site is not known.

carried out at a sample temperature of ∼50 K maintained by a closed-cycle helium cryostat. 1,3,5-Tris(4-mercaptophenyl)benzene (TMB) was thermally evaporated from a home-built Knudsen cell with crucible temperatures around 145 °C. During deposition and STM imaging the substrate was held at room temperature. STM images were acquired at room temperature and processed by line-wise leveling and 3 × 3 Gaussian filtering.

■ RESULTS AND DISCUSSION

In this work, adatom-mediated coordination of TMB (cf. Figure 1a) on Cu(111) into two-dimensional (2D) metal–organic networks based on thiolate–copper coordination bonds is presented. Bulk synthesis already yielded copper–thiolate metal–organic frameworks (MOF)[3] and linear polymeric structures.[4] While surface-confined coordination networks based on copper–carboxylate coordination bonds have been reported by several groups,[1d,2b,5] to our knowledge, this type of interlinking chemistry has not been utilized for surface-supported 2D systems. We demonstrate that upon thermal annealing an initial precursor structure is converted into copper–thiolate coordinated networks mediated by the free-adatom gas of the Cu(111) surface. Interestingly, the thiolate–copper complexes found in this study contain copper dimers which coordinate two thiolates. In bulk MOFs, interconnects typically consist of single metal atoms as coordination centers. In the proposed system the coordinating metal dimers are additionally stabilized by adsorption on the surface, possibly rendering them unique for surface-confined systems.

In a first preparation step TMB is deposited by thermal sublimation under ultrahigh-vacuum (UHV) conditions on Cu(111) at room temperature and characterized by means of in situ STM and LEED. Figure 1b depicts an STM topograph of the resulting self-assembled densely packed trigonal structure with an STM-derived lattice parameter of (1.30 ± 0.05) nm. Accompanying LEED measurements (cf. Supporting Information) aid to identify the monolayer as a commensurate $3\sqrt{3} \times 3\sqrt{3}$ R30° superstructure. Upon adsorption on reactive metal surfaces thiols deprotonate and become thiolates which are anchored by sulfur–metal bonds.[6] Both the size of the unit cell and the three-fold symmetric appearance of adsorbed TMB in STM topographs substantiate the assumption that TMB fully deprotonates on Cu(111) and forms three covalent S–Cu bonds with the substrate; a tentative model of the precursor structure is depicted in Figure 1c. However, the formation of a densely packed monolayer indicates a non-negligible contribution from intermolecular interactions for structure formation. While the azimuthal orientation of TMB within the unit cell and with respect to the substrate can be inferred from STM topographs, its absolute adsorption position with respect to the copper substrate remains unknown. Interestingly, the TMB-derived trithiolate molecule is also commensurate with the substrate; i.e. for its actual azimuthal orientation all sulfur atoms reside on similar adsorption sites and can thus simultaneously optimize their interaction with the substrate. Covalent anchoring by three peripheral sulfur groups stabilizes a planar adsorption geometry of TMB on Cu(111), whereas monothiolates tend to adsorb upright,[6d,7] or inclined, as it is the case for halogen-substituted thiophenols.[8]

In a second preparation step, thermal annealing at 160–200 °C for ∼10 min converts the self-assembled trithiolate monolayers into two polymorphs which are both identified as metal–organic coordination networks. STM topographs of both metal-coordinated polymorphs are depicted in Figure 2. Conversion of the initial precursor structure into 2D metal-coordinated networks is accompanied by substantial reorientation and repositioning of TMB molecules, but most importantly by introducing intermolecular copper–thiolate coordination bonds.

The effect of annealing is 2-fold: First lateral mobility of the trithiolate species is enhanced. Second, the area density of the free copper adatom gas is greatly increased, whereby a sufficient amount of highly mobile coordination centers is supplied.

The influence of a 2D adatom gas has been recognized as an important contribution to the surface chemistry of metals.[9] The adatom gas originates from a temperature-dependent evaporation/condensation equilibrium at step-edges. At lower temperatures, processes with lower activation energies are dominant, i.e. mass transport along step-edges. For higher temperatures mass exchange between step-edges and terraces is the dominant process[10] leading to a drastic increase of the free adatom concentration already at moderate temperatures of ∼500 K.[11] Conversion of initially intermolecular hydrogen-bonded networks into metal–organic networks on Cu(111) in this temperature range have similarly been reported by Matena[2b]

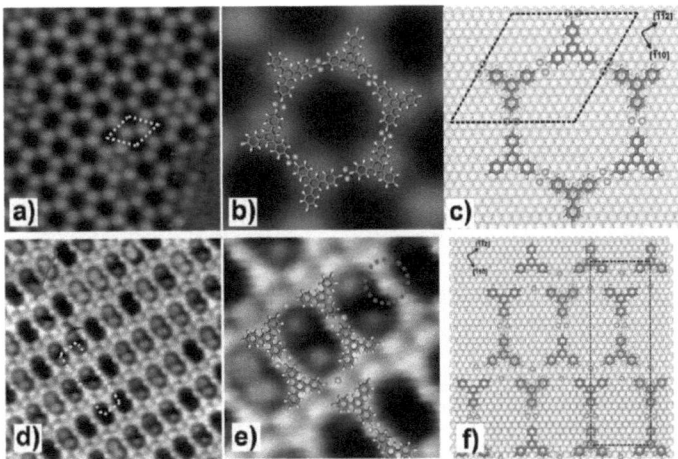

Figure 2. (a) STM topograph of honeycomb structure with unit cell indicated ($U_T = -1.0$ V, $I_T = 67$ pA, 24×24 nm^2, $a = b = 3.4$ nm, $\gamma = 120°$) and (b) close-up (6.5×6.5 nm^2) of honeycomb structure with molecular model. (c) Tentative model of the honeycomb structure including the substrate; the hexagonal unit cell is indicated by black dashed lines. (d) STM topograph of dimer row structure with unit cell indicated ($U_T = -0.8$ V, $I_T = 121$ pA, 18×18 nm^2, $a = 2.2$ nm, $b = 6.6$ nm, $\gamma = 90°$) and (e) close-up (6.9 nm $\times$ 6.9 nm^2) of dimer row structure with molecular model. Protruding features are observed in the STM contrast at the center of a dimer (marked by the dashed circle) and hint toward metal coordination. (f) Tentative model of the dimer row structure including the substrate; the rectangular unit cell is indicated by black dashed lines.

and Pawin.[2a] In the latter work, annealing at lower temperatures leads to a partially hydrogen-bonded and partially metal-coordinated polymorph. As already mentioned above for TMB, two different metal-coordinated networks, a hexagonal honeycomb and a centered rectangular dimer row structure emerge. In both structures copper adatoms coordinate TMB molecules via their thiolate groups.

The hexagonal honeycomb structure (Figure 2a−c) has a lattice parameter of 3.4 nm and belongs to the plane space group p6mm. The second structure is less symmetric (c2mm) and is composed of rows of dumbbell-shaped dimers (Figure 2e−f). Adjacent dimer rows are offset by exactly half a lattice parameter exactly along the row axes, resulting in a rectangular centered nonprimitive unit cell. During the conversion of the densely packed precursor structure into the porous networks, entrapment of excess TMB molecules within the pores occurs frequently and gives rise to additional contrast features within the pores as evident from Figure 2. This observation is in accord with other experiments on periodic and irregular porous surface-supported networks where molecules, either deposited in excess, captured during structure formation,[12,13] or additionally deposited[14], were likewise trapped within the pores.

The honeycomb and dimer row structure were observed in coexistence as shown in Figure 3. The relative ratio of both phases slightly depends on the initial coverage of the precursor structure, with a preference for the more densely packed dimer row structure at higher coverages. It is noteworthy that the morphology of the two metal-coordinated TMB polymorphs resembles those of rubrene monolayers both on (110) and (111) copper surfaces[15] and Au(111).[16] Yet, despite the similar appearance of rubrene vs TMB monolayers in STM topographs, their

Figure 3. Overview STM topograph illustrating the coexistence of both phases ($U_T = -1.64$ V, $I_T = 66$ pA, 116×116 nm^2) on Cu(111). Red arrows indicate first occurrence of degraded molecules starting to appear at annealing temperatures around 220 °C. White arrows indicate crystallographic directions of the substrate.

self-assembly is notably different. In the gas phase rubrene is highly nonplanar and even remains nonplanar after adsorption. TMB, on the other hand, is slightly nonplanar in the gas phase due to an out-of-plane rotation of its peripheral mercaptophenyl units with respect to the phenyl ring at the center. However, the

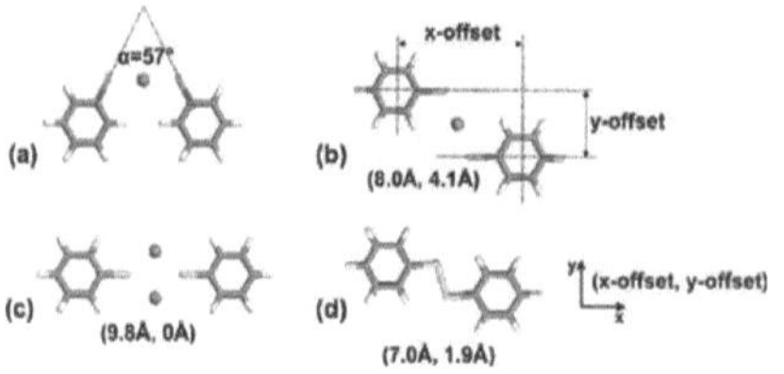

Figure 4. DFT geometry optimized intermolecular bonding schemes for interconnected phenylthiolates via (a) one-center trans-coordination syn-conformation ($\alpha = 57°$), (b) one-center trans-coordination anti-conformation (x-offset = 8.0 Å, y-offset = 4.1 Å), (c) two-center coordination bond (x-offset = 9.8 Å, no y-offset), (d) covalent coupling via disulfur-bridge (x-offset = 7.0 Å, y-offset = 1.9 Å). For all bonds, the center-to-center distance of the phenyl groups (x-offset) and the perpendicular axial offset (y-offset) respectively are given in parentheses.

mirror symmetric STM appearance of TMB in all observed structures suggests that it becomes planar upon adsorption due to interactions with the substrate. Also the pronounced chirality of rubrene affects its self-assembly and leads to the expression of chiral structures and aggregates,[17] while no indication of chirality is discernible in any of the TMB structures. Lastly, the interactions which drive self-assembly are distinctly different for rubrene and TMB. In the precursor structure of TMB, covalent bonds between sulfur and copper play an important role, whereas after the phase transition metal coordination becomes the predominant interaction. On the other hand, these types of interactions are absent in rubrene self-assembly, where van der Waals interactions, higher multipole electrostatic interactions, and substrate-mediated interactions govern structure formation.[18]

For a fundamental understanding of the thiolate−copper coordination bonds, DFT calculations (cf. Supporting Information for details) have been performed of the connecting nodes, modeled by two phenylthiolates and corresponding copper centers. Four different intermolecular bonding schemes were considered: metal-coordination bonds mediated by one or two copper atoms, and a covalent disulfur bridge. Motivated by experimental results on thiolate−gold complexes[19] both syn−trans and anti−trans arrangements were taken into account for the one-center coordination bonds. To reduce the computational cost only the outer phenylthiolate parts were simulated, and for ease of calculations the explicit substrate influence has been neglected in this first approximation. Typically, intermolecular bond lengths of adsorbed systems are altered in comparison to the gas phase, but these differences are normally small and, especially for large entities, often below the accuracy of STM measurements. DFT results on these simplified model systems are depicted in Figure 4. Major findings can be summarized as follows. Only two-center coordination bonds facilitate linear interconnection. One-center coordination bonds and covalent disulfur bridges both result in lateral offsets perpendicular to the bond axis.

The one-center anti conformational arrangement and the disulfur bridge result in a lateral bond offset of ∼4.1 Å and ∼1.9 Å, respectively. These lateral offsets are sufficiently large to be identifiable in STM topographs. In the one-center syn conformational assembly the molecular axes include an angle of 57°, which is significantly larger than the 35° reported for Au-coordinated methylthiolate by Voznyy and co-workers.[19] This can readily be explained by steric repulsion, being more pronounced for the bulky phenyl ligands as compared to that for methyl groups.

According to DFT results, the total binding energy of two-center coordination complexes is strongest with a value of 555 kJ/mol. Total binding energies of trans−syn (394 kJ/mol) and trans−anti (397 kJ/mol) one-center coordination complexes are comparable, but notably lower than for the two-center complex. The covalent disulfur bridge is the weakest bond with a strength of 151 kJ/mol. Covalent S−S bonds exhibit a certain variability of bond angles and energies,[20] with the latter value being within the typical range.

Both the honeycomb and the dimer row structure contain dumbbell shaped TMB dimers as basic structural motif. High resolution STM images of both polymorphs occasionally show protruding features between dimers; an example is marked in Figure 2e. An additional STM topograph of the honeycomb structure very clearly showing protruding contrast features at the center position between adjacent molecules is provided in the Supporting Information (cf. Figure S2). Although these contrast features hint toward intermolecular bonds through metal-coordination, they do not allow to unambiguously infer the exact number of metal-coordination centers per bond. Yet, since no lateral displacement occurs along the dimer axis, the DFT calculations suggest coordination of the thiolate groups by two copper atoms as depicted in Figure 4c. This conclusion is further substantiated by comparison of experimental and theoretical bond lengths. STM data yield a center-to-center distance of 2.0 (±0.2) nm between two TMB molecules in the dimer. DFT results in combination with the intramolecular distance between central and outer phenyl rings in TMB (0.46 nm for a geometry optimized isolated molecule) postulate a dimer center-to-center distance of 1.90 nm. Accordingly, the experimental lattice parameter of the honeycomb structure (3.4 nm) is in good agreement with the anticipated lattice parameter from the two-center coordination interconnect (3.3 nm), while both the one-center coordination scenario (2.9 nm) and the covalent disulfur bridge (2.8 nm) would yield notably smaller lattice parameters. From the p6mm symmetry it can be concluded that all intermolecular bonds in the honeycomb structure are equivalent. A complete model of the honeycomb structure is overlaid in the STM topograph depicted in Figure 2b, and a tentative structural model including the substrate is separately presented in Figure 2c.

Yet, due to its lower symmetry the dimer row structure cannot consistently be explained solely on the basis of two-center coordination bonds. Thus, a bonding scheme is proposed involving different types of intermolecular bonds. Both appearance and center-to-center distance of dimers (∼2.0 nm) within the rows are similar to the honeycomb structure. Also in both structures the dimers have the same orientation with respect to the Cu(111) substrate, i.e. their axes are oriented along the $\overline{112}$ direction. This registry points toward a distinct epitaxial relation between the dimers and the Cu(111) substrate. Hence, it is concluded that also the dimers within the dimer row structure are similarly interconnected by two-center coordination bonds. These dimers assemble in parallel rows, where adjacent rows are shifted half a lattice parameter with respect to the row direction. STM topographs exhibit protruding contrast features directly above and below intrarow neighbors (marked in Figure 2d by a red arrow) which point toward a trans−syn arrangement, as implemented in the models in e and f of Figure 2, where the former is overlaid to the STM image and the latter model includes the substrate. Since coordinating copper atoms cannot be resolved separately, as is also the case for the two atoms coordinating the dimers, dimer−dimer coordination by more than one Cu atom cannot be excluded. Polynuclear copper−thiolate coordination bonds are common,[21] where coordinating copper clusters are further stabilized by cuprophilicity.[22] For instance, coordination by Cu_3 clusters has also been observed previously for bulk systems.[3]

Since the doubly copper-coordinated dimer is the structural motif of both the honeycomb and the dimer row structure, its adsorption geometry on the Cu(111) substrate is of particular interest. A tentative model which takes both the STM-derived orientation with respect to the substrate and the DFT-derived atomic configuration of the interconnect into account is provided in Figure 5. This model illustrates that both coordinating copper atoms could indeed adsorb at similar sites, for instance as tentatively shown in three-fold hollow sites. The actual distances between the two coordinating copper atoms in the proposed surface-confined metal-coordination complexes range most likely between the two extreme values, i.e. 2.87 Å as derived from our DFT calculations for a fully neglected and 2.55 Å—the Cu(111) lattice parameter—for a fully effective substrate influence. Therefore, in summary the actual Cu−Cu distance might result from a compromise and interplay between the periodic potential of the substrate which favors the smaller Cu−Cu spacing closer to the lattice constant and the gas phase Cu−Cu spacing which is more influenced by the orbital configuration and hybridization of the thiolate and the copper centers.

In order to reveal the specific role of the Cu(111) substrate for the phase transition and to shed light on the chemical properties and reactivity of copper adatoms, the same type of experiments with similar sample preparation protocols were conducted on Ag(111). Room-temperature deposition of TMB on Ag(111) also yields a densely packed trigonal structure (cf. Figure 6a, $a = 1.35 \pm 0.05$ nm,). Since the lattice parameter of the trigonal structure on Ag(111) is similar to the value obtained on Cu(111) within experimental error, the precursor structure on Ag(111) is also identified as densely packed deprotonated trithiolates which are covalently anchored to the substrate through three peripheral thiolate−metal bonds. While the initial TMB precursor structures appear similar on both substrates, the response to thermal annealing is distinctly different on Ag(111): annealing up to 250 °C for ~1 h did not result in a phase transition; however, after the sample annealed at 300 °C for ~1 h, disordered glassy networks were observed in coexistence with remnants of the precursor structure; a typical STM topograph is depicted in Figure 6b. Nonperiodic glassy organic networks have recently gained substantial interest. Particularly nice examples include metal-coordination networks of nonlinear, prochiral ditopic organic linkers with cobalt atoms[23] and hydrogen-bonded cytosine networks on Au(111).[13b,24] Further, structurally comparable irregular organic networks have also been observed, when halogenated precursor molecules were polymerized by surface-mediated reactions into covalent networks.[25] While the irregularity of the glassy metal-coordination and hydrogen-bonded networks arises from the low symmetry of the building blocks in combination with the energetic equivalence of various different basic intermolecular bond motifs, the degree of disorder typically

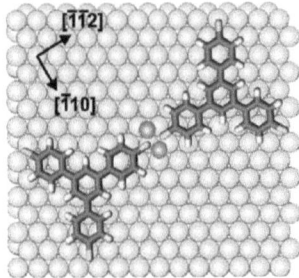

Figure 5. Tentative model of the adsorption geometry of a doubly copper coordinated TMB dimer including the Cu(111) substrate. Both coordinating copper atoms (colored red) could adsorb on similar lattice sites to simultaneously optimize their adsorption energy, e.g. as shown on three-fold hollow sites (yellow: substrate atoms).

Figure 6. STM topographs of TMB deposited on Ag(111) (a) as-deposited at room temperature ($U_T = -0.33$ V, $I_T = 44$ pA, 30×39 nm^2) and (b) after annealing to 300 °C for ~1 h ($U_T = -0.98$ V, $I_T = 85$ pA, 30×30 nm^2). The precursor structure on Ag(111) appears similar to Cu(111) with a similar lattice parameter of (1.35 ± 0.05) nm. Annealing of as-deposited samples at conditions which would already result in irreversible deterioration of the networks on Cu(111), yields a glassy disordered network on Ag(111). Within the disordered networks interconnected dimers can be discerned (marked by white arrows); their dimensions are consistent with formation of covalent disulfur bridges.

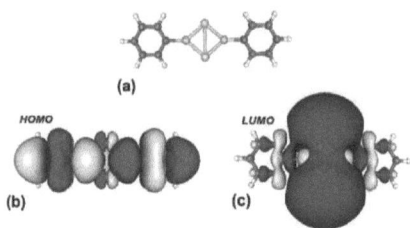

Figure 7. DFT results of structure and frontier molecular orbitals of a two-center coordination bond interconnecting two phenylthiolates (a) geometry, (b) HOMO, (c) LUMO of the complex.

observed for covalent networks is owed to the irreversibility of the covalent molecular interlinks under the growth conditions, which inhibits postcorrection of topological defects. A closer look at the disordered TMB structures on Ag(111) reveals dimers with a distinct lateral offset to the interconnecting axis; examples are marked in Figure 6b by white arrows. The lateral offset is consistent with DFT-derived values for disulfur bridges; consequently, we propose that the dimers observed after a thermal treatment of the precursor structure on Ag(111) are covalently interlinked.

An interesting, not entirely solved question is why adatom-mediated formation of metal-coordination networks was observed for the same TMB molecules on Cu(111) but not on Ag(111). On both substrates, the densely packed precursor structures obtained upon room temperature deposition are structurally quite similar, and differences in their precise epitaxial relations between both substrates can hardly account for the absence of metal-coordination networks on Ag(111). Also the temperature dependent densities of the adatom gases on both surfaces are comparable[26] ruling out adatom availability as a decisive criterion. Even more so, since tempering the Ag(111) samples up to significantly higher temperatures should have provided a sufficient amount of adatoms. Nevertheless, silver coordinated TMB metal—organic networks have never been observed, but only irregular networks. The absence of metal-coordination networks on Ag(111) is best explained by a different affinity of Cu vs Ag adatoms to form metal-coordination bonds with thiolates. This hypothesis is in accord with experimental findings on the adatom-mediated formation of carboxylate-based metal-coordination networks on Cu(111) vs Ag(111). While metal-coordination networks of trimesic acids were readily observed on Cu(111), their formation was again absent on Ag(111).[9] Interestingly, preceding deposition of copper onto Ag(111) promoted the formation of trimesic acid metal-coordination networks also on this substrate. These results are consistently explained by the assumption that copper deposition on Ag(111) introduces a copper adatom gas which is in equilibrium with the deposited copper islands. The higher reactivity of this artificially introduced, extrinsic copper adatom gas is the driving force for formation of copper metal-coordination networks.

Moreover, the present results on covalently interlinked TMB molecules on Ag(111) indicate that the lateral offset of disulfur-bridged molecules can clearly be resolved in STM and hence indirectly prove that the porous TMB networks on Cu(111) are not built up by disulfur-bridged molecules.

In order to illustrate the electronic properties of the copper—thiolate metal-coordination interlink, DFT derived frontier molecular orbitals are depicted in Figure 7. Evidently, both HOMO and LUMO of bicoordinated phenylthiolates exhibit intensity at the bond site. The LUMO appears to be more localized at the bond, whereas the HOMO is evenly distributed across the metal-coordination complex. Such delocalization allows for coherent electron transport through the metal-coordination bond, rendering this interconnection chemistry a suitable candidate for interlinking single molecules into more complex molecular electronics circuitry.

■ SUMMARY

In summary, adatom-mediated 2D metal—organic networks were synthesized on Cu(111) by thermal annealing of a self-assembled precursor structure. The two observed metal—organic networks are based on metal-coordination bonds between thiolates and either one or two copper adatoms. Comparison between DFT-derived and experimental bond lengths and geometries aided in the identification of intermolecular coordination bonds. In contrast, deposition on Ag(111) resulted in a similar precursor structure, but annealing at higher temperatures only resulted in irregular structures, where monomers are interconnected by covalent disulfur bridges. These pronounced differences between both surfaces are explained by a higher affinity of copper adatoms as compared to silver adatoms to form metal-coordination bonds with thiolates.

As suggested by the spatial distributions of their frontier molecular orbitals, copper—thiolate complexes are fully electronically conjugated. This intriguing feature renders copper—thiolate coordination bonds particularly interesting for organic conductors and molecular electronics. Especially the envisioned molecular electronics applications not only require precise electronic function within a single molecule but also equally directional communication between molecules. Yet, up to now the issue of interconnecting single-molecule devices into more complex circuits is not satisfactorily addressed. Numerous studies concluded that, for contacts and interconnects, bond topology on the atomic level is of utmost importance due to the coherent nature of electron transport in molecular electronics. Hence, means to interconnect molecular entities in an atomically defined manner without perturbing or interrupting electronic conjugation are urgently required. Thiol groups in combination with copper coordination centers are ideally suited as "solder" for molecular electronics due to their electronic conjugation, their relatively high stability, and not at least due to their compatibility with self-assembly bottom-up fabrication techniques.

■ ASSOCIATED CONTENT

ⓈSupporting Information. Synthesis and calculational details, additional LEED and STM results. This material is available free of charge via the Internet at http://pubs.acs.org.

■ AUTHOR INFORMATION

Corresponding Author
markus@lackinger.org

Present Address
#Institut National de la Recherche Scientifique, Université du Québec, 1650 boulevard Lionel-Boulet, Varennes, QC, J3X 1S2 Canada.

■ ACKNOWLEDGMENT

Financial support by the Deutsche Forschungsgemeinschaft (DFG) within the Nanosystems-Initiative Munich (NIM) and with FOR 516 (Siegen) is gratefully acknowledged. St.S. and G.E. are particularly grateful for support by the Elitenetzwerk Bayern and the Hanns-Seidel Stiftung. T.S. acknowledges support by the Fonds der Chemischen Industrie.

■ REFERENCES

(1) (a) Barth, J. V. *Annu. Rev. Phys. Chem.* **2007**, *58*, 375–407. (b) Stepanow, S.; Lin, N.; Barth, J. V. *J. Phys.: Condens. Matter* **2008**, *20*, 184002. (c) Stepanow, S.; Lingenfelder, M.; Dmitriev, A.; Spillmann, H.; Delvigne, E.; Lin, N.; Deng, X. B.; Cai, C. Z.; Barth, J. V.; Kern, K. *Nat. Mater.* **2004**, *3*, 229–233. (d) Barth, J. V.; Weckesser, J.; Lin, N.; Dmitriev, A.; Kern, K. *Appl. Phys. A* **2003**, *76*, 645–652. (e) Ruben, M.; Rojo, J.; Romero-Salguero, F. J.; Uppadine, L. H.; Lehn, J. M. *Angew. Chem., Int. Ed.* **2004**, *43*, 3644–3662.

(2) (a) Pawin, G.; Wong, K. L.; Kim, D.; Sun, D. Z.; Bartels, L.; Hong, S.; Rahman, T. S.; Carp, R.; Marsella, M. *Angew. Chem., Int. Ed.* **2008**, *47*, 8442–8445. (b) Matena, M.; Stöhr, M.; Riehm, T.; Bjork, J.; Martens, S.; Dyer, M. S.; Persson, M.; Lobo-Checa, J.; Müller, K.; Enache, M.; Wadepohl, H.; Zegenhagen, J.; Jung, T. A.; Gade, L. H. *Chem.—Eur. J.* **2010**, *16*, 2079–2091.

(3) He, J.; Yang, C.; Xu, Z. T.; Zeller, M.; Hunter, A. D.; Lin, J. H. *J. Solid State Chem.* **2009**, *182*, 1821–1826.

(4) Che, C. M.; Li, C. H.; Chui, S. S. Y.; Roy, V. A. L.; Low, K. H. *Chem.—Eur. J.* **2008**, *14*, 2965–2975.

(5) (a) Lin, N.; Dmitriev, A.; Weckesser, J.; Barth, J. V.; Kern, K. *Angew. Chem., Int. Ed.* **2002**, *41*, 4779–4783. (b) Perry, C. C.; Haq, S.; Frederick, B. G.; Richardson, N. V. *Surf. Sci.* **1998**, *409*, 512–520. (c) Dougherty, D. B.; Maksymovych, P.; Yates, J. T. *Surf. Sci.* **2006**, *600*, 4484–4491.

(6) (a) Driver, S. M.; Woodruff, D. P. *Surf. Sci.* **2000**, *457*, 11–23. (b) Ferral, A.; Paredes-Olivera, P.; Macagno, V. A.; Patrito, E. M. *Surf. Sci.* **2003**, *525*, 85–99. (c) Keller, H.; Simak, P.; Schrepp, W.; Dembowski, J. *Thin Solid Films* **1994**, *244*, 799–805. (d) Konopka, M.; Turansky, R.; Dubecky, M.; Marx, D.; Stich, I. *J. Phys. Chem. C* **2009**, *113*, 8878–8887. (e) Maksymovych, P.; Sorescu, D. C.; Yates, J. T. *Phys. Rev. Lett.* **2006**, *97*, 146103. (f) Sardar, S. A.; Syed, J. A.; Ikenaga, E.; Yagi, S.; Sekitani, T.; Wada, S.; Taniguchi, M.; Tanaka, K. *Nucl. Instrum. Methods Phys. Res., Sect. B* **2003**, *199*, 240–243.

(7) Di Castro, V.; Bussolotti, F.; Mariani, C. *Surf. Sci.* **2005**, *598*, 218–225.

(8) Wong, K. L.; Lin, X.; Kwon, K. Y.; Pawin, G.; Rao, B. V.; Liu, A.; Bartels, L.; Stolbov, S.; Rahman, T. S. *Langmuir* **2004**, *20*, 10928–10934.

(9) Lin, N.; Payer, D.; Dmitriev, A.; Strunskus, T.; Woll, C.; Barth, J. V.; Kern, K. *Angew. Chem., Int. Ed.* **2005**, *44*, 1488–1491.

(10) Giesen, M. *Surf. Sci.* **1999**, *442*, 543–549.

(11) Giesen, M. *Prog. Surf. Sci.* **2001**, *68*, 1–153.

(12) Griessl, S.; Lackinger, M.; Edelwirth, M.; Hietschold, M.; Heckl, W. M. *Single Mol.* **2002**, *3*, 25–31.

(13) (a) Ruben, M.; Payer, D.; Landa, A.; Comisso, A.; Gattinoni, C.; Lin, N.; Collin, J. P.; Sauvage, J. P.; De Vita, A.; Kern, K. *J. Am. Chem. Soc.* **2006**, *128*, 15644–15651. (b) Otero, R.; Lukas, M.; Kelly, R. E. A.; Xu, W.; Laegsgaard, E.; Stensgaard, I.; Kantorovich, L. N.; Besenbacher, F. *Science* **2008**, *319*, 312–315. (c) Kühne, D.; Klappenberger, F.; Decker, R.; Schlickum, U.; Brune, H.; Klyatskaya, S.; Ruben, M.; Barth, J. V. *J. Phys. Chem. C* **2009**, *113*, 17851–17859.

(14) (a) Stöhr, M.; Wahl, M.; C.H., G.; Riehm, T.; Jung, T. A.; Gade, L. H. *Angew. Chem., Int. Ed.* **2005**, *44*, 7394–7398. (b) Stöhr, M.; Wahl, M.; Spillmann, H.; Gade, L. H.; Jung, T. A. *Small* **2007**, *3*, 1336–1340.

(15) (a) Miwa, J. A.; Cicoira, F.; Bedwani, S.; Lipton-Duffin, J.; Perepichka, D. F.; Rochefort, A.; Rosei, F. *J. Phys. Chem. C* **2008**, *112*, 10214–10221. (b) Miwa, J. A.; Cicoira, F.; Lipton-Duffin, J.; Perepichka, D. F.; Santato, C.; Rosei, F. *Nanotechnology* **2008**, *19*, 424021.

(16) Pivetta, M.; Blüm, M.-C.; Patthey, F.; Schneider, W.-D. *Angew. Chem., Int. Ed.* **2008**, *47*, 1076–1079.

(17) Blüm, M.-C.; Cavar, E.; Pivetta, M.; Patthey, F.; Schneider, W.-D. *Angew. Chem., Int. Ed.* **2005**, *44*, 5334–5337.

(18) Tomba, G.; Stengel, M.; Schneider, W.-D.; Baldereschi, A.; De Vita, A. *ACS Nano* **2010**, *4*, 7545–7551.

(19) Voznyy, O.; Dubowski, J. J.; Yates, J. T.; Maksymovych, P. *Angew. Chem., Int. Ed.* **2009**, *131*, 12989–12993.

(20) Steudel, R. *Angew. Chem., Int. Ed. Engl.* **1975**, *14*, 655–664.

(21) Ahte, P.; Palumaa, P.; Tamm, T. *J. Phys. Chem. A* **2009**, *113*, 9157–9164.

(22) Hermann, H. L.; Boche, G.; Schwerdtfeger, P. *Chem.—Eur. J.* **2001**, *7*, 5333–5342.

(23) Marschall, M.; Reichert, J.; Weber-Bargioni, A.; Seufert, K.; Auwärter, W.; Klyatskaya, S.; Zoppellaro, G.; Ruben, M.; Barth, J. V. *Nature Chem.* **2010**, *2*, 131–137.

(24) Kelly, R. E. A.; Lukas, M.; Kantorovich, L. N.; Otero, R.; Xu, W.; Mura, M.; Laegsgaard, E.; Stensgaard, I.; Besenbacher, F. *J. Chem. Phys.* **2008**, *129*, 184707.

(25) (a) Gutzler, R.; Walch, H.; Eder, G.; Kloft, S.; Heckl, W. M.; Lackinger, M. *Chem. Commun.* **2009**, 4456–4458. (b) Walch, H.; Gutzler, R.; Sirtl, T.; Eder, G.; Lackinger, M. *J. Phys. Chem. C* **2010**, *114*, 12604–12609.

(26) Zhang, J. M.; Song, X. L.; Zhang, X. J.; Xu, K. W.; Ji, V. *Surf. Sci.* **2006**, *600*, 1277–1282.

7.5 Incorporation dynamics of molecular guests into two-dimensional supramolecular host networks at the liquid-solid interface

Georg Eder, Stephan Kloft, Natalia Martsinovich, Kingsuk Mahata, Michael Schmittel, Wolfgang M. Heckl, and Markus Lackinger

Langmuir, **2011**, 27 (22), 13563–13571

http://dx.doi.org/10.1021/la203054k

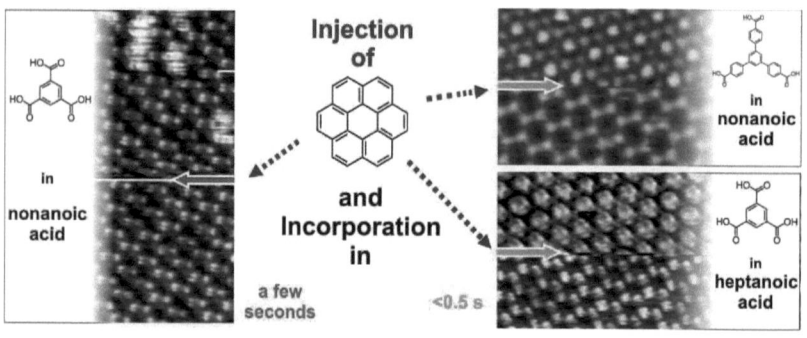

Reprinted with permission from [92]. Copyright 2011 American Chemical Society

Langmuir

ARTICLE

pubs.acs.org/Langmuir

Incorporation Dynamics of Molecular Guests into Two-Dimensional Supramolecular Host Networks at the Liquid−Solid Interface

Georg Eder,[†,‡] Stephan Kloft,[‡] Natalia Martsinovich,[§] Kingsuk Mahata,[∥] Michael Schmittel,[∥] Wolfgang M. Heckl,[†,⊥] and Markus Lackinger*,[†,‡,⊥]

[†]TUM School of Education and Center for NanoScience (CeNS), Tech. Univ. Munich, Schellingstrasse 33, 80799 Munich, Germany
[‡]Department for Earth and Environmental Sciences, Ludwig-Maximilians-University, Theresienstrasse 41, 80333 Munich, Germany
[§]Department of Chemistry, University of Warwick, Gibbet Hill Rd CV4 7AL, Coventry, U.K.
[∥]Organische Chemie I, University of Siegen, Adolf-Reichwein-Strasse 2, 57068 Siegen, Germany
[⊥]Deutsches Museum, Museumsinsel 1, 80538 Munich, Germany

Supporting Information

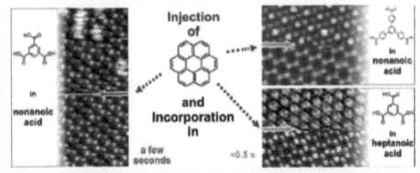

ABSTRACT: The objective of this work is to study both the dynamics and mechanisms of guest incorporation into the pores of 2D supramolecular host networks at the liquid−solid interface. This was accomplished by adding molecular guests to prefabricated self-assembled porous monolayers and the simultaneous acquisition of scanning tunneling microscopy (STM) topographs. The incorporation of the same guest molecule (coronene) into two different host networks was compared, where the pores of the networks either featured a perfect geometric match with the guest (for trimesic acid host networks) or were substantially larger than the guest species (for benzenetribenzoic acid host networks). Even the moderate temporal resolution of standard STM experiments in combination with a novel injection system was sufficient to reveal clear differences in the incorporation dynamics in the two different host networks. Further experiments were aimed at identifying a possible solvent influence. The interpretation of the results is aided by molecular mechanics (MM) and molecular dynamics (MD) simulations.

■ INTRODUCTION

Two-dimensional supramolecular porous host networks have attracted substantial interest, not the least because of their ability to incorporate molecular guests, as often directly verified and visualized by high-resolution scanning probe microscopy.[1] Two-dimensional host networks are commonly surface-supported and feature a periodic arrangement of identical nanometer-sized pores bordered and defined by network-forming molecules. Generally, porous structures oppose nature's tendency toward dense packing by virtue of relatively strong directional interactions such as hydrogen bonds,[2] metal coordination bonds,[3] or even alkyl chain interdigitation.[4] Some networks exhibit permanent porosity, whereas for several initially densely packed systems it has been shown that the self-assembly of porous polymorphs can be favored either by low solute concentrations[5] or by a templating effect, where guest inclusion renders occupied porous polymorphs thermodynamically more stable.[6,7] The size, shape, and spatial arrangement of pores in 2D host networks are predominantly controlled by the structure and functionalization of the constituting molecules; however, they can also depend on various parameters such as the temperature,[8] the type of solvent,[9,10] the concentration,[5,11] the stoichiometry in heteromeric networks,[12] and the substrate.[13]

Homomeric and heteromeric supramolecular host monolayers may form both under ultrahigh vacuum (UHV) conditions and at the liquid−solid interface. Host networks have proven to be efficient templates for controlling the coadsorption of molecular guests with subnanometer precision.[5,14] Whereas in a typical UHV experiment molecular guests are incorporated into the host networks by codeposition,[15−17] at the liquid−solid interface guests are often added to the supernatant liquid phase. This was demonstrated for numerous examples (e.g., isophthalic acid derivatives and p-phenylene vinylenes reported by de Feyter et al.[18] or tetracarboxylic acid tectons[19]). In thermodynamically equilibrated systems, guest incorporation is observed only when the associated enthalpic gain exceeds the entropic loss.

In many cases, planar polycyclic aromatic hydrocarbons (PAH)[20,21] were introduced as guests, for instance, coronene[6] or hexabenzocoronene.[22] Nonplanar compounds such as C₆₀ fullerenes are also suitable as guests[23] and have recently been shown to template bilayer growth effectively.[17] Very often, molecular guests were geometrically matched to the host pores.[24] However, the

Received: August 5, 2011
Revised: September 20, 2011
Published: September 27, 2011

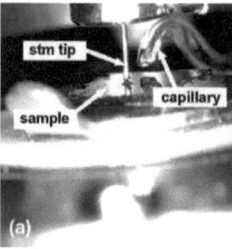 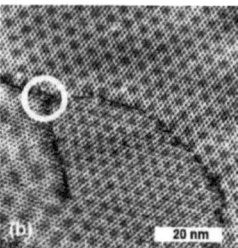 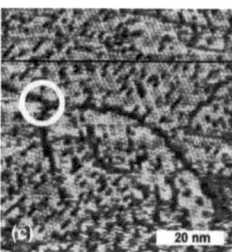

Figure 1. (a) Photograph of the STM setup showing the additional bent glass capillary used for injection. STM topographs (frame time = 66.9 s, V_t = +0.67 V, I_t = 81 pA) demonstrate the position stability upon the injection of additional solution. (b) Large-scale STM topograph of a TMA monolayer in **9A** including domain boundaries, marked with a white circle. Three different Moiré patterns with different periodicities can be recognized. This is caused by the occurrence of various epitaxial relationships between the adsorbate and substrate lattices and indicates a weak substrate influence. (c) STM topograph acquired ∼140 s after the injection of additional solution containing COR guest molecules. The cross-correlation of both STM topographs resulted in a lateral offset of ∼2 nm.

inclusion of more than one guest molecule of the same species in one pore[2] and even heteromeric supramolecular aggregates as guests have also been reported.[25] Moreover, host networks may be responsive to the guest molecules and thus adjust their pore sizes accordingly.[26] In addition to the structural characterization of host−guest networks, temperature-[22] and concentration-induced[5] adsorption and desorption, lateral[27] and vertical[14] manipulation, interpore guest diffusion,[22] and size and shape dependences of their adsorption[28] have been studied.

In many cases, scanning tunneling microscopy (STM) was used to characterize the structures of both the initially empty host and the occupied host−guest networks with submolecular resolution. However, because of the limited temporal resolution of STM and the lack of complementary analytical techniques at the liquid−solid interface, little is known about the dynamics and mechanisms of guest incorporation.

Here we present a first approach toward revealing the incorporation dynamics of molecular guests into supramolecular host networks at the liquid−solid interface. For this purpose, high-resolution STM image acquisition was continued while a solution containing the molecular guest was added to the supernatant liquid phase above a prefabricated host network. In this study, we compare the incorporation of the same molecular guest into two different host networks. In the first case, the geometry of a single guest molecule perfectly matches the supramolecular pore. In the second case, however, the host pores are substantially larger than the guest molecules. We demonstrate for the chosen systems and the proposed experimental approach that even the moderate temporal resolution of standard STM experiments is sufficient to reveal clear differences between incorporation into the two different host networks. First, the experimental setup that allows the simultaneous imaging and addition of guest molecules to the solution is described. Second, the results for guest incorporation into two different hexagonal host networks with different pore sizes are presented. The discussion and interpretation of the results is aided by molecular mechanics (MM) and molecular dynamics (MD) simulations.

■ EXPERIMENTAL SECTION

All experiments were conducted with a home-built, versatile, drift-stable SPM operated by ASC 500 control electronics from attocube Systems AG. All STM measurements were performed at room temperature (25 ± 2 °C) with mechanically cut Pt/Ir (90/10) tips. The initial host networks were prepared by applying a droplet (2.5 μL) of the respective saturated solution onto the basal plane of freshly cleaved, highly oriented pyrolytic graphite (HOPG, grade ZYB, Optigraph GmbH Berlin, 6 × 6 mm²). STM topographs were acquired directly at the liquid−solid interface with the tip immersed in the solution in constant-current mode with tip voltages in the range of +0.5 to +1.0 V and current set points of around 70 pA.

Supramolecular host networks were prepared from saturated solutions of either 1,3,5-benzenetricarboxylic acid (trimesic acid, TMA, $C_9H_6O_6$, Sigma-Aldrich) or 1,3,5-benzenetribenzoic acid (BTB, $C_{27}H_{18}O_6$, synthesized following a known literature procedure[29]) in either heptanoic acid (**7A**, Sigma-Aldrich) or nonanoic acid (**9A**, Sigma-Aldrich). Coronene (COR, $C_{24}H_{12}$, Sigma-Aldrich) was used as a guest molecule for both host networks and likewise was dissolved in the respective fatty acid as a solvent. The solubility limit of COR in **9A**, as determined by UV−vis absorption spectroscopy, was reached at a concentration of 1.0 ± 0.2 mmol/L. Because the initially deposited solution of either TMA or BTB did not contain any guest molecules, the COR concentration of the injected solution became diluted. The volumes of initially present and additionally applied solution were similar, thus the final guest concentration was diluted to 50%, neglecting solvent losses.

■ COMPUTATIONAL DETAILS

Molecular mechanics (MM) and molecular dynamics (MD) simulations were performed with the Tinker program[30] using the MM3 force field[31,32] that was modified[33] to describe double hydrogen bonds in carboxylic acid dimers. Because of cooperative resonance effects,[34,35] double hydrogen bonds (e.g., in carboxylic acid or amide dimers) are stronger than equivalent single hydrogen bonds, whereas standard force fields, such as the widely used MM3, neglect these resonance effects and thus underestimate the strength of cyclic double hydrogen bonds. To consider these cooperative effects and correctly describe the hydrogen bonding in MM and MD simulations, the MM3 force field was modified according to a previous study.[33]

TMA and BTB networks were modeled with a TMA or BTB hexamer adsorbed on a hydrogen-terminated graphene sheet. Different orientations and positions of the host networks with respect to the graphite surface were simulated, yet without qualitatively and quantitatively affecting the outcome. This underlines the weak interaction with the graphite substrate and the low corrugation of the surface potential for comparatively large molecules.

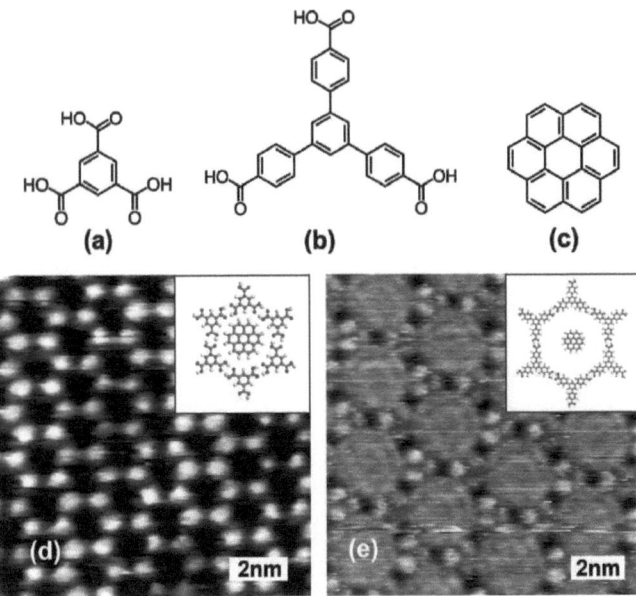

Figure 2. Chemical structures of (a) trimesic acid (TMA, $C_9H_6O_6$), (b) 1,3,5-benzenetribenzoic acid (BTB, $C_{27}H_{18}O_6$), and (c) coronene (COR, $C_{24}H_{12}$). STM topographs of pure host networks resulting from the self-assembly of (d) TMA (V_t = +0.86 V, I_t = 57 pA) and (e) BTB (V_t = +0.48 V, I_t = 48 pA). The insets illustrate the corresponding host−guest structures and exemplify the size relation of the six-membered pore in comparison to the COR guest.

MD simulations were performed using the NVT ensemble with a Berendsen thermostat at temperatures of 298 and 400 K, with an integration time step of 0.1 fs. The z coordinates of the graphite substrate were fixed during MD simulations; all coordinates of two corner carbon atoms of the substrate and two O atoms of the TMA or BTB hexamer were also fixed to prevent rotation. Because of the complexity of the problem and computational limitations, the solvent was not considered in our calculations.

■ RESULTS AND DISCUSSION

In the first step, host networks were prepared and imaged by STM at the liquid−solid interface. Once the STM was drift-stable and a submolecular resolution of the host network was obtained, an additional droplet of solution containing the guest molecule (∼2.5 μL) was added via an injection system with a bent glass capillary aimed at the sample (Figure 1a). The guest solutions were injected under visual camera control by means of a thoroughly mechanically decoupled syringe outside the enclosure of the STM. The injection neither impaired the high resolution nor caused substantial drifts. The performance of the injection system was verified by adding a guest-containing solution to the liquid phase above the host network while questioning the positional stability of an unambiguous structural feature in subsequent scans. A sample area with domain boundaries was chosen, and the lateral offset between two scans before (Figure 1b) and ∼140 s after (Figure 1c) the injection was measured. Although thermal drift also contributes to any lateral offset, the encircled domain boundary was shifted by less than 2 nm. Hence, the proposed injection system is well suited for the study of dynamic effects after the addition of further solutions, possibly containing different types of molecules, with minimal drift.

Utilizing the proposed method, the dynamics of coronene (COR) incorporation was characterized for trimesic acid (TMA) host networks in both **7A** and **9A**, for benzenetribenzoic acid (BTB) host networks in **9A**. BTB in **7A** was not studied because as a result of solvent-induced polymorphism BTB in **7A** self-assembles into a densely packed polymorph that is not suitable for guest incorporation.[10] Molecular structures are shown in Figure 2a−c. COR, a planar PAH, was chosen as a guest species because of its favorable interaction with the graphite substrate and its perfect geometrical match with TMA pores.[14] Owing to this superior match, COR molecules are stabilized and immobilized in TMA pores and can thus be imaged with submolecular resolution.

At first, the self-assembly of the respective host networks was initiated by depositing a droplet of saturated BTB or TMA solution onto the substrate and verified by STM imaging. Representative STM topographs and corresponding structural models of both host networks are depicted in Figure 2d,e. Under the chosen experimental conditions (concentration, solvent, and

Figure 3. Incorporation of COR in TMA host networks in 9A. (a) STM topograph of the prefabricated TMA host network in 9A. The horizontal arrow and t_0 mark the position and time, respectively, of the guest injection. (b) STM topograph illustrating the transient intermediate adsorption state of the guest molecules. The insets show detailed views of COR guests in this intermediate state and in the final state. The TMA host network is indicated by light-gray triangles. (c) STM topograph of the final state, where guest molecules were incorporated into the TMA monolayer and submolecular resolution was again obtained. (V_t = +0.73 V, I_t = 71 pA, and T = 297.0 K for all topographs).

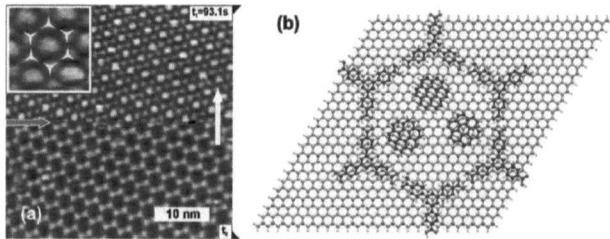

Figure 4. (a) STM topograph of COR incorporation into BTB host networks in 9A (V_t = +0.76 V, I_t = 51 pA, and T = 297 K). The slow-scan direction is marked by the vertical white arrow, and the horizontal arrow indicates the scan line where saturated COR solution was added; also visible is a slight disturbance. The intrapore contrast changes within one to three scan lines (i.e., in less than 0.5 s). The inset represents a detailed view of an occupied pore, and the six bordering BTB molecules are indicated by light-gray triangles. (b) Geometry-optimized MM simulation of three COR molecules within one BTB cavity. Although up to four COR molecules would fit into one BTB cavity, MD simulations indicate that this situation is unstable and one COR desorbs. However, because of the lateral mobility of the loosely packed three remaining COR molecules within the BTB pore, molecular resolution cannot be obtained. Thus, it is not possible to verify the postulated structure experimentally.

temperature), both TMA and BTB form porous, topologically similar honeycomb networks: both tricarboxylic acids adsorb in a planar arrangement on the graphite surface, and all three carboxylic groups of each molecule are engaged in cyclic double hydrogen bonds in a self-complementary manner. The plane symmetry group of both analogous monolayer structures is the same, p6mm, whereas the TMA and BTB networks differ substantially in the lattice parameter (1.7 nm for TMA vs 3.2 nm for BTB) and accordingly in pore size (∼1.0 nm for TMA vs ∼2.8 nm for BTB). Two different types of submolecular BTB contrast were regularly observed. A single BTB molecule appeared as three clearly separated protrusions in a 3-fold arrangement as in Figure 2e. In this case, the intermolecular hydrogen bonds appear even lower than the pore. In the other, more frequently observed case, BTB appears as a smeared-out feature that represents the contour of the molecule; an example of this contrast is shown in Figure 4a. The occurrence of either contrast was not found to depend systematically on the bias voltage and was attributed to the imaging properties of the specific tip.

To study the dynamics of COR incorporation, the solution containing the guests was added to the liquid phase during STM image acquisition by means of the injection system described above. The experimental results are depicted in Figure 3 (TMA in 9A), Figure S1 (TMA in 7A), and Figure 4 (BTB in 9A). In the first part of the first STM topographs, the pure host networks were imaged and then the guest-containing solutions were added at the scan lines marked by horizontal arrows. Vertical arrows denote the slow-scan direction. The number of additionally supplied COR molecules is sufficient to occupy all cavities (i.e., 90 COR molecules per TMA cavity and 370 COR molecules per BTB cavity) when saturated COR solution in 9A was injected.

Clear differences are evident between COR incorporation into TMA and BTB host networks in 9A. The larger BTB pores feature very fast COR incorporation within the time resolution of the experiment (Figure 4a), whereas for TMA host networks, the guests appear rather fuzzy and streaky in the STM image directly after the injection. Because the host network can still be imaged with unimpaired resolution, we interpret the initial fuzzy appearance of COR not as a loss of resolution but as a transient intermediate adsorption state. Albeit already centered above TMA cavities, in STM images COR guests initially appear to be considerably wider than anticipated from their geometric dimensions. See the inset in Figure 3b for a detailed view. Moreover, COR appears to be fuzzy and striped in the STM images of this intermediate adsorption state, and submolecular details remain unresolved. However, without the need to change the scan parameters or improve the STM tip, the contrast changed over time to the known appearance of binary COR + TMA networks as corroborated by at least 25 independent experimental runs. See the inset in Figure 3c for a detailed view. Submolecular details of the guest species were then resolved, indicating the immobilization of COR within the TMA pores. The TMA host network itself was not affected by COR incorporation (i.e., the lattice parameter, orientation, and lateral extension of the domains remained similar within experimental error).

Distinct differences in the incorporation dynamics were also noticed for different COR concentrations. The equilibrium pore occupation depends on the total guest concentration in the liquid phase. For lower COR concentrations, not all pores become occupied anymore. A statistical analysis of the spatial distribution of occupied pores reveals a non-random distribution of guests, where incorporated COR molecules tend to cluster. The experimentally determined average number of occupied adjacent pores for each guest was significantly higher than expected for a random distribution. However, at this point it remains unclear how the attractive interaction between guests is mediated.

Also, the incorporation rate, as observed by simultaneous STM imaging, was highly dependent on the COR concentration in the injected solution. An experimental series with different COR concentrations and the TMA network reveals a significantly lower COR incorporation rate for lower concentrations (Supporting Information, Figure S2). This is also confirmed for BTB networks, where a low guest concentration induces a time delay until the first COR molecules arrive at the interface, while the incorporation itself takes place equally fast (Supporting Information, Figure S3).

This is a consequence of the concentration-dependent arrival rate of the guest species at the interface, rendering COR incorporation a diffusion-limited process for low guest concentrations in the additional solution. To arrive at the interface, in the proposed experiments guest molecules need to diffuse through the liquid film that is still present from the previous self-assembly of the initial host networks. For lower COR concentrations in the additionally applied solution, the driving force for diffusion (i.e., the concentration gradient) is smaller; consequently, more time is required for COR molecules to diffuse to the interface. However, even for very low COR concentrations the intermediate adsorption state has always been observed for TMA networks, thereby excluding the COR diffusion rate as a possible origin.

The incorporation of COR guests into the significantly larger pores of the BTB host network proceeded entirely differently. In contrast to the TMA networks where a fuzzy, streaky intermediate state was observed, the incorporation of COR molecules into the BTB host network from saturated solution was observed in less than 0.5 s (i.e., rapidly within the temporal resolution of the experiment). After this very fast incorporation as compared to TMA, the intrapore contrast did not change anymore over time, indicating that equilibrium had already been reached. In contrast to COR + TMA, occupied BTB pores appeared with protruding featureless internal contrast, and a submolecular resolution of incorporated guests could not be obtained. Nevertheless, a clear difference in the apparent heights between initially empty and occupied pores indicates the incorporation of COR guests. BTB pores are substantially larger than TMA pores, and there is no distinct geometric match between COR guests and BTB pores. From the STM data, the exact number of COR guest molecules adsorbed in each BTB pore cannot be inferred. Because the BTB network was imaged with submolecular resolution at the same time, the absence of internal STM contrast within the pore is attributed to physical reasons rather than to technical difficulties. In the more spacious BTB pore, the residual mobility (i.e., the insufficient immobilization of COR guests) impairs high-resolution STM.[36] At high guest concentrations, it appears likely that as many COR molecules as can possibly fit into each BTB pore are incorporated. According to MM simulations, each BTB cavity can accommodate up to four COR molecules in planar adsorption geometry.

To get a better impression of the residual mobility of COR guests within BTB pores, MD simulations were conducted at a simulation temperature of 298 K. The MM-optimized geometries of four planar adsorbed COR molecules in one BTB pore served as initial structures for MD simulations, whereby different runs all yielded the same result: one COR molecule is ejected from the BTB cavity, and the three remaining COR guests oscillate around their preferred position. (See Supporting Information Figures S5 and S6 for representations of the COR movement and a detailed analysis of the trajectories.) We rationalize this surprising result in terms of both enthalpy and entropy. For three COR molecules in the BTB pore, the MM-derived adsorption energy per COR molecule is 161.5 kJ/mol, whereas for four COR molecules, the adsorption energy per molecule is slightly smaller, 156.1 kJ/mol. Thus, four COR molecules in the pore feature a much larger adsorption energy per pore; however, in terms of the enthalpy per molecule, it will be more favorable to fill all pores with three COR molecules each rather than to fill some pores with four CORs and leave other pores empty. The second factor that has to be taken into account is entropy. The desorption of the fourth COR molecule from the pore brings about additional degrees of freedom, associated with both the increased mobility of the three COR molecules remaining in the pore and with the translational and rotational entropy gained by the desorption of one COR molecule in solution. Both favorable entropic contributions render the occupation of the BTB pores by three COR molecules thermodynamically more stable, especially at higher temperatures.

The easy mobility of the three COR molecules that remain adsorbed in the BTB pore explains the featureless intrapore contrast in STM topographs. From STM experiments and MD simulations, we conclude that the lateral immobilization of COR guests is less effective in BTB pores than in the perfectly size-matched TMA cavities. Because the intrapore STM contrast of occupied BTB host−guest networks did not change over time, we conclude that equilibrium is reached in less than 0.5 s. For the injection of diluted COR solution (~4% saturation), again no intermediate state was observed but direct incorporation was observed (Supporting Information, Figure S3). All occupied cavities appeared with similar contrast, hinting at uniform occupation by the same number of COR molecules. The main difference for lower COR concentrations was again a time delay between injection and incorporation, underlining the rate-limiting influence of solution diffusion.

The interpretation of the experimental results is complemented by MM and MD simulations in order to elucidate the nature of the intermediate state as observed for TMA networks in 9A. First, MM was used to quantify the energetics of COR incorporation in both networks and identify possible energy barriers for the incorporation process. Geometry-optimized structures of a single COR molecule in a TMA pore and three COR molecules in a

BTB pore are depicted in Figures 2d and 4b. The adsorption energy of COR in the TMA network accounts for 183.7 kJ/mol, which is larger than the average value of 161.5 kJ/mol per molecule in the case of three COR molecules in the BTB network.

Besides the perfect geometric match of COR in TMA pores, incorporated guest molecules are additionally stabilized by 18 hydrogen bonds with the pore wall (6 C−H···O$_{carbonyl}$ and 12 C−H···O$_{hydroxyl}$ hydrogen bonds, see Supporting Information, Figure S4). These hydrogen bonds give rise to a preferred azimuthal orientation of COR molecules within TMA pores and account for the high rotational barrier of ∼10.0 kJ/mol. This barrier height should facilitate thermally activated rotational hopping at room temperature at an extremely high rate, when standard values of 10^{13}–10^{15} s^{-1} are assumed for the exponential prefactor. However, high-resolution STM topographs of COR immobilized in a TMA network clearly exhibit nonrotationally symmetric submolecular contrast features of the guest.[14] To bring these apparently contradicting results together, it is assumed that the rotational hopping event occurs extremely fast as compared to the dwell time of the COR guest in one of the six rotationally symmetric potential minima. Consequently, the time-averaging STM contrast shows apparently immobile COR. The energy of the COR−TMA hydrogen bonds can be estimated by comparing the adsorption energy of COR in the TMA pore (which includes both COR−TMA and COR−graphite binding) and on pristine graphite. The former energy is 39.7 kJ/mol larger than the latter; therefore, each hydrogen bond on average contributes 2.2 kJ/mol to the total binding energy. The typical energy range of C−H···O hydrogen bonds is 4−8 kJ/mol,[37] which is considerably higher. However, the associated O···H distances of these stronger hydrogen bonds lie around 2.5 Å, whereas for COR in TMA the respective distance is around 2.74 Å. Here, the hydrogen bond length is determined by the geometric relation between COR guests and TMA pores, and the larger distance results in a weaker hydrogen bond.

Despite this strong directional host−guest interaction, the structure of the host network is not affected. Pores are neither widened nor narrowed upon COR incorporation as verified by MM of a single six-membered TMA pores on graphite. The distance between the centers of diametrically opposite TMA molecules before COR incorporation amounts to 1.86 nm and does not change significantly through COR inclusion. The extension or contraction of pores in host networks due to guest incorporation would also influence adjacent pores and thus alter their hosting properties.

To identify possible energy barriers for COR incorporation into TMA pores, the energy of an azimuthally and horizontally aligned guest molecule as a function of its vertical distance to the graphite surface was evaluated by MM calculations. Figure 5 compares the energies of COR above TMA on graphite and above pristine graphite; the difference between both energy curves represents the additional stabilization of COR through the TMA network, which already becomes effective ∼0.5 nm above the equilibrium position. The geometry of the host pore was allowed to adjust at each step, but the changes in the TMA atoms' positions were smaller than 0.075 Å compared to the coordinates of the unfilled TMA hexamer. The energy curve in Figure 5 does not hint at any energy barriers for incorporation, and the equilibrium position corresponds to a distance of 0.34 nm between the graphite and COR planes.

In summary, MM simulations show that the COR + TMA match is perfect in the sense that the TMA pores do not relax

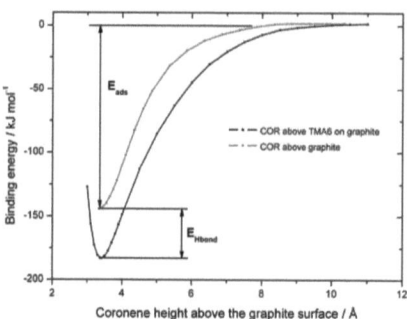

Figure 5. MM-derived adsorption energies of COR above pristine graphite (red curve) and above a 6-fold TMA pore on graphite (black curve) as a function of the vertical distance. The distance-dependent energy difference arises from the additional stabilization of COR due to the TMA host network.

upon guest incorporation and there is a strong host−guest interaction via hydrogen bonds. However, no barrier was found that could explain the slow incorporation of COR into the TMA network.

Next, we consider another possibility for the intermediate adsorption state. MM revealed a metastable adsorption site of COR directly atop TMA with a relatively large adsorption energy of 74.5 kJ/mol. The geometry-optimized structure is depicted in Figure 6b. Therefore, we used MD simulations to explore the possibility that the metastable adsorption of COR atop TMA can account for the experimentally observed intermediate state.

At a simulation temperature of 298 K, COR remains adsorbed at the metastable adsorption site for the simulated time span of 0.4 ns. The corresponding trajectory of the center of mass is shown in Figure 7a. Extended simulated time spans of up to 10 ns (using a longer integration time step, i.e., 1.0 fs instead of 0.1 fs) yielded qualitatively similar results. However, when the simulation temperature is increased to 400 K, COR migrates from the metastable adsorption site to the stable adsorption configuration in the pore within the first 70 ps. The corresponding center-of-mass trajectory is depicted in Figure 7b. Because the solvent is disregarded in the presented MD calculations, a higher simulation temperature might offer a more realistic description of the system: the increased simulation temperature enhances the adsorbate mobility and facilitates desorption, which may mimic the effect of the solvent−solute interaction. Overall, MD simulations indicate that the metastable adsorption of COR atop TMA may persist during the short time spans accessible by MD calculations at room temperature.

Although the MD simulations already offer the entrapment of COR in a metastable adsorption state as a possible explanation, the timescales of the calculations and experiment differ by 10 orders of magnitude. To study the influence of other parameters, additional experiments were conducted in another solvent. The self-assembly of TMA from saturated **7A** solutions also yields the honeycomb network,[38] and similar incorporation experiments were conducted with COR dissolved in **7A**. In contrast to incorporation experiments of COR with TMA in **9A**, in most experimental runs with short-chain-length solvent **7A** no long-lived

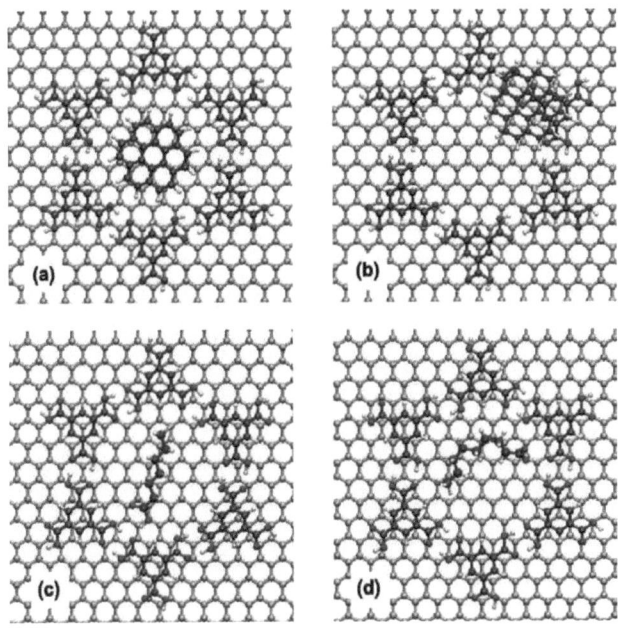

Figure 6. MM-optimized adsorption geometries of COR and solvent molecules within a six-membered TMA pore. (a) Structure of incorporated COR. The guest is additionally stabilized by 18 intermolecular hydrogen bonds. (b) Structure of metastable COR adsorption directly atop TMA with a calculated adsorption energy of 74.5 kJ/mol. Representative adsorption configurations of (c) a single 7A and (d) a single 9A solvent molecule in a TMA pore.

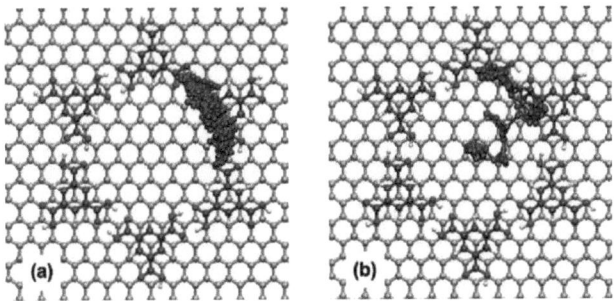

Figure 7. MD simulations of COR and the TMA network. COR was initially adsorbed in the metastable position atop TMA, and the time evolution was simulated (a) for a temperature of 298 K and a time span of 0.4 ns and (b) for a temperature of 400 K and a time span of 0.3 ns. Dark-brown dots represent the center-of-mass trajectory of COR molecules.

intermediate state was observed (Supporting Information, Figure S1). Instead, COR incorporation occurred relatively fast, thus the results with the TMA network in 7A are qualitatively similar to COR incorporation into the larger BTB pores. Because equally fast incorporation without an observable intermediate adsorption state has never been seen for TMA in **9A**, these additional experiments in **7A** indicate a solvent influence on the COR incorporation dynamics.

It is commonly assumed that at the liquid−solid interface host network pores are not actually empty but occupied by solvent or even additional solute molecules.[39] However, the stabilization energy of incorporated solvent molecules is inferior because of their smaller size and constraints imposed by the pore geometry, resulting in a dynamic equilibrium featuring constant exchange. STM topographs represent only the time average of this dynamic equilibrium, and clear contrast features that would allow one to infer the adsorption geometry of solvent molecules are normally not observed. The geometric relation between pore and solvent molecules is exemplified by the MM-optimized geometries for the coadsorption of single **7A** and **9A** molecules within TMA pores in Figure 6c,d. Although the TMA pores are too small to accommodate a fully extended **9A** solvent molecule, a curled up **9A** solvent molecule fits into the TMA pore (Figure 6d) and can gain additional stabilization from interaction with the pore walls. Several fully extended **9A** molecules conveniently fit into BTB pores; however, their packing still remains rather loose. However, from temperature-programmed desorption (TPD) experiments it is known that the adsorption energies of alkanes and fatty acids on graphite increase linearly with the length of the aliphatic tail.[40] Although these results were obtained from well-ordered monolayers, where fatty acids dimerize via hydrogen bonds, they will also qualitatively hold true for their coadsorption in pores of host networks as demonstrated below by MM calculations.

Calculated adsorption energies of **9A** in TMA pores are up to 100.8 kJ/mol (for the most stable adsorption configurations found involving hydrogen bonding to TMA molecules), whereas **7A** adsorbs somewhat less strongly (up to 94.6 kJ/mol): a good geometric fit of **7A** with TMA pores is counterbalanced by the shorter chain length and consequently fewer interactions with the graphite surface. In the larger BTB pores, two types of adsorption sites and corresponding adsorption energies can be found: (i) on the edges of the pores, where solvent molecules can form hydrogen bonds with BTB, the adsorption energies are up to 104.7 and 87.4 kJ/mol for **9A** and **7A**, respectively; (ii) in the centers of the pores, the adsorption energies are similar to those found on the pristine graphite surface: up to 70.7 and 57.3 kJ/mol for **9A** and **7A**, respectively. Thus, the solvent molecules positioned in the center of the BTB pores would desorb more easily. In any case, our MM calculations confirm that the adsorption energies of **7A** are always smaller than for **9A**. Consequently, desorption rates of the shorter-chain-length fatty acids are larger, giving rise to a more dynamic exchange with the liquid phase for **7A** as compared to that for **9A**.

To explain the pore-size and solvent dependence of COR incorporation, we propose that the lifetime of the intermediate COR adsorption state mainly depends on the time required to displace solvent molecules that are coadsorbed in the pore. An additional influence due to the exact geometric match of COR and TMA pores, which gives rise to an entropic barrier for incorporation, is indicated by MD simulations. However, because the intermediate state is absent for **7A**, we conclude that this entropic barrier cannot account for the experimentally observed lifetime.

According to MM simulations, the average stabilization energy of **9A** in TMA pores is larger than that of **7A**. However, the stabilization energies of both solvent molecules in smaller TMA pores are larger than for BTB pores, especially for adsorption in the center of the pore. Consequently, solvent desorption as a necessary preceding event for COR incorporation takes place on a much faster timescale for the larger BTB pore. Moreover, for smaller TMA pores, solvent desorption can become sterically impaired by COR adsorption in the vicinity of the pore either in an intermediate state atop TMA or even above the solvent-filled pore. Hindered solvent desorption, however, can be excluded for the larger BTB pores because a single COR molecule covers only a fraction of the pore area and leaves enough space for solvent molecules to desorb. In summary, the higher stabilization energy of **9A** solvent molecules and hindered solvent desorption in smaller TMA pores, possibly assisted by a high entropic barrier due to the exact geometric match, can explain the emergence of an intermediate adsorption state for COR incorporation into TMA pores in **9A**. The lower stabilization energy of **7A** in TMA pores reduces the time required to displace previously adsorbed solvent molecules with COR molecules to such an extent that the intermediate adsorption state cannot be observed anymore in the STM experiment.

■ CONCLUSION AND OUTLOOK

For the first time, the incorporation dynamics of molecular guests into 2D host networks is studied at the liquid−solid interface. By means of continued STM imaging, while a guest-containing solution was added to the supernatant liquid phase above a prefabricated supramolecular host network, a transient intermediate state of COR incorporation into TMA networks could be identified. Interestingly, incorporation was observed in less than 0.5 s in similar experiments for the BTB host network whose pores are significantly larger than COR guests. For the TMA network, the solvent dependence was also studied, where the intermediate state was always observed in **9A** but only very rarely in **7A**. MM and MD simulations were applied to reveal the nature of the intermediate incorporation state. MM indicates a perfect match between COR and TMA pores. Hydrogen bonds additionally stabilize the guest molecule, but the structure of the host pore is not affected by guest incorporation. Structural simulations by MM also show that up to four COR molecules can be accommodated in the larger BTB pore, but MD simulations suggest stable adsorption of three COR molecules only. In any case, because of the loose fit and the residual mobility of the guests within the pore it was not possible to obtain molecular resolution of the guests. MM simulations do not hint toward any incorporation barrier for COR in the TMA network but indicate a metastable adsorption position of COR atop TMA. The time evolution of this metastable adsorption was further studied by MD. At room temperature, metastable adsorption persists for the accessible time spans, whereas for slightly elevated temperatures, incorporation takes place in less than 1 ns. Although these MD simulations offer a possible origin of the intermediate state, a distinct experimentally observed solvent dependence suggests an alternative explanation: The pores of host networks at the liquid−solid interface are not empty but are occupied by solvent molecules, although clear adsorption geometries cannot be inferred from STM measurements because of the highly dynamic nature of solvent coadsorption. However, for guest incorporation, the solvent molecules within the pores need to be displaced and the dynamics of this process is much faster for **7A** than for **9A** and for larger BTB pores than for smaller TMA pores.

The proposed experimental technique allows the monitoring of dynamic processes at liquid−solid interfaces on the molecular level. We show that STM imaging can be continued even with submolecular resolution when additional solution is added to an already present liquid phase. In the current study, this additional solution provides new molecules as guests for coadsorption in host networks, but the experimental technique could also be used to change concentrations or mix solvents in order to study the effect on the interfacial monolayer directly. Albeit the temporal resolution in these experiments is not extraordinary high, the method bears the potential to provide fundamental insights into monolayer dynamics at the liquid−solid interface, thus contributing to the understanding of kinetic processes in supramolecular self-assembly.

■ ASSOCIATED CONTENT

ⓈSupporting Information. Additional STM topographs and MM and MD simulations. This material is available free of charge via the Internet at http://pubs.acs.org.

■ AUTHOR INFORMATION

Corresponding Author
*E-mail: markus@lackinger.org.

■ ACKNOWLEDGMENT

Financial support by the Nanosystems-Initiative Munich (NIM) and the Bayerische Forschungsstiftung is gratefully acknowledged. G.E. is particularly grateful for the financial support of the Hanns-Seidel-Stiftung. M.S. is indebted to the DFG (FOR 516).

■ REFERENCES

(1) Kudernac, T.; Lei, S. B.; Elemans, J. A. A. W.; De Feyter, S. *Chem. Soc. Rev.* **2009**, *38*, 402.
(2) MacLeod, J. M.; Ivasenko, O.; Fu, C.; Taerum, T.; Rosei, F.; Perepichka, D. F. *J. Am. Chem. Soc.* **2009**, *131*, 16844.
(3) Stepanow, S.; Lingenfelder, M.; Dmitriev, A.; Spillmann, H.; Delvigne, E.; Lin, N.; Deng, X.; Cai, C.; Barth, J. V.; Kern, K. *Nat. Mater.* **2004**, *3*, 229.
(4) Tahara, K.; Furukawa, S.; Uji-I, H.; Uchino, T.; Ichikawa, T.; Zhang, J.; Mamdouh, W.; Sonoda, M.; De Schryver, F. C.; De Feyter, S.; Tobe, Y. *J. Am. Chem. Soc.* **2006**, *128*, 16613.
(5) Lei, S. B.; Tahara, K.; De Schryver, F. C.; Van der Auweraer, M.; Tobe, Y.; De Feyter, S. *Angew. Chem., Int. Ed.* **2008**, *47*, 2964.
(6) Furukawa, S.; Tahara, K.; De Schryver, F. C.; Van der Auweraer, M.; Tobe, Y.; De Feyter, S. *Angew. Chem., Int. Ed.* **2007**, *46*, 2831.
(7) Blunt, M.; Lin, X.; Gimenez-Lopez, M. D.; Schroder, M.; Champness, N. R.; Beton, P. H. *Chem. Commun.* **2008**, 2304.
(8) Gutzler, R.; Sirtl, T.; Dienstmaier, J. F.; Mahata, K.; Heckl, W. M.; Schmittel, M.; Lackinger, M. *J. Am. Chem. Soc.* **2010**, *132*, 5084.
(9) Liu, J.; Zhang, X.; Yan, H. J.; Wang, D.; Wang, J. Y.; Pei, J.; Wan, L. J. *Langmuir* **2010**, *26*, 8195.
(10) Kampschulte, L.; Lackinger, M.; Maier, A. K.; Kishore, R. S. K.; Griessl, S.; Schmittel, M.; Heckl, W. M. *J. Phys. Chem. B* **2006**, *110*, 10829.
(11) Meier, C.; Landfester, K.; Künzel, D.; Markert, T.; Groß, A.; Ziener, U. *Angew. Chem., Int. Ed.* **2008**, *47*, 3821.
(12) Kampschulte, L.; Werblowsky, T. L.; Kishore, R. S. K.; Schmittel, M.; Heckl, W. M.; Lackinger, M. *J. Am. Chem. Soc.* **2008**, *130*, 8502.
(13) Liang, H.; He, Y.; Ye, Y. C.; Xu, X. G.; Cheng, F.; Sun, W.; Shao, X.; Wang, Y. F.; Li, J. L.; Wu, K. *Coord. Chem. Rev.* **2009**, *253*, 2959.
(14) Griessl, S. J. H.; Lackinger, M.; Jamitzky, F.; Markert, T.; Hietschold, M.; Heckl, W. M. *Langmuir* **2004**, *20*, 9403.
(15) Theobald, J. A.; Oxtoby, N. S.; Phillips, M. A.; Champness, N. R.; Beton, P. H. *Nature* **2003**, *424*, 1029.
(16) Forrest, S. R. *Chem. Rev.* **1997**, *97*, 1793.
(17) Blunt, M. O.; Russell, J. C.; Gimenez-Lopez, M. D.; Taleb, N.; Lin, X. L.; Schroder, M.; Champness, N. R.; Beton, P. H. *Nat. Chem.* **2011**, *3*, 74.
(18) De Feyter, S.; De Schryver, F. C. *J. Phys. Chem. B* **2005**, *109*, 4290.
(19) Champness, N. R.; Slater, A. G.; Beton, P. H. *Chem. Sci.* **2011**, *2*, 1440.
(20) Wasserfallen, D.; Fischbach, I.; Chebotareva, N.; Kastler, M.; Pisula, W.; Jackel, F.; Watson, M. D.; Schnell, I.; Rabe, J. P.; Spiess, H. W.; Mullen, K. *Adv. Funct. Mater.* **2005**, *15*, 1585.
(21) Müllen, K.; Rabe, J. P. *Acc. Chem. Res.* **2008**, *41*, 511.
(22) Schull, G.; Douillard, L.; Fiorini-Debuisschert, C.; Charra, F.; Mathevet, F.; Kreher, D.; Attias, A. J. *Nano Lett.* **2006**, *6*, 1360.
(23) Perepichka, D. F.; MacLeod, J. M.; Ivasenko, O.; Rosei, F. *Nanotechnology* **2007**, 18.
(24) Adisoejoso, J.; Tahara, K.; Okuhata, S.; Lei, S.; Tobe, Y.; De Feyter, S. *Angew. Chem., Int. Ed.* **2009**, *48*, 7353.
(25) Lei, S.; Surin, M.; Tahara, K.; Adisoejoso, J.; Lazzaroni, R.; Tobe, Y.; Feyter, S. D. *Nano Lett.* **2008**, *8*, 2541.
(26) Zhang, X.; Chen, T.; Yan, H.-J.; Wang, D.; Fan, Q.-H.; Wan, L.-J.; Ghosh, K.; Yang, H.-B.; Stang, P. J. *ACS Nano* **2010**, *4*, 5685.
(27) Griessl, S. J. H.; Lackinger, M.; Jamitzky, F.; Markert, T.; Hietschold, M.; Heckl, W. M. *J. Phys. Chem. B* **2004**, *108*, 11556.
(28) Schull, G.; Douillard, L.; Fiorini-Debuisschert, C.; Charra, F.; Mathevet, F.; Kreher, D.; Attias, A. J. *Adv. Mater.* **2006**, *18*, 2954.
(29) Weber, E.; Hecker, M.; Koepp, E.; Orlia, W.; Czugler, M.; Csoregh, I. *J. Chem. Soc., Perkin Trans. 2* **1988**, 1251.
(30) Ponder, J. W.; Richards, F. M. *J. Comput. Chem.* **1987**, *8*, 1016.
(31) Lii, J. H.; Allinger, N. L. *J. Comput. Chem.* **1998**, *19*, 1001.
(32) Allinger, N. L.; Yuh, Y. H.; Lii, J. H. *J. Am. Chem. Soc.* **1989**, *111*, 8551.
(33) Martsinovich, N.; Troisi, A. *J. Phys. Chem. C* **2010**, *114*, 4376.
(34) Gilli, P.; Bertolasi, V.; Ferretti, V.; Gilli, G. *J. Am. Chem. Soc.* **1994**, *116*, 909.
(35) Beck, J. F.; Mo, Y. *J. Comput. Chem.* **2007**, *28*, 455.
(36) Lei, S.; Tahara, K.; Feng, X.; Furukawa, S.; De Schryver, F. C.; Müllen, K.; Tobe, Y.; De Feyter, S. *J. Am. Chem. Soc.* **2008**, *130*, 7119.
(37) Dienstmaier, J. F.; Mahata, K.; Walch, H.; Heckl, W. M.; Schmittel, M.; Lackinger, M. *Langmuir* **2010**, *26*, 10708.
(38) Lackinger, M.; Griessl, S.; Heckl, W. A.; Hietschold, M.; Flynn, G. W. *Langmuir* **2005**, *21*, 4984.
(39) Tahara, K.; Okuhata, S.; Adisoejoso, J.; Lei, S. B.; Fujita, T.; De Feyter, S.; Tobe, Y. *J. Am. Chem. Soc.* **2009**, *131*, 17583.
(40) Müller, T.; Flynn, G. W.; Mathauser, A. T.; Teplyakov, A. V. *Langmuir* **2003**, *19*, 2812.

PUBLICATION

7.6 Solution preparation of two dimensional covalently linked networks by polymerization of 1,3,5-tri(4-iodophenyl)benzene on Au(111)

Georg Eder, Emily F. Smith, Wolfgang M. Heckl, Peter H. Beton, and Markus Lackinger

submitted, **2012**

http://dx.doi.org/

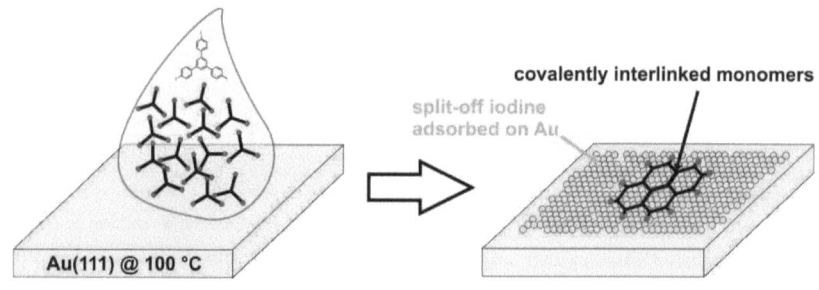

Reprinted with permission from [258]. Copyright 2013 American Chemical Society

Solution Preparation of Two-Dimensional Covalently Linked Networks by Polymerization of 1,3,5-Tri(4-iodophenyl)benzene on Au(111)

Georg Eder,[†] Emily F. Smith,[‡] Izabela Cebula,[‡,§] Wolfgang M. Heckl,[†,∥] Peter H. Beton,[§,*] and Markus Lackinger[†,∥,*]

[†]TUM School of Education and Center for NanoScience (CeNS), Tech Univ Munich, Schellingstrasse 33, 80799 Munich, Germany, [‡]School of Chemistry and [§]School of Physics and Astronomy, University of Nottingham, University Park, Nottingham NG7 2RD, United Kingdom, and [∥]Deutsches Museum, Museumsinsel 1, 80538 Munich, Germany

ABSTRACT The polymerization of 1,3,5-tri(4-iodophenyl)-benzene (TIPB) on Au(111) through covalent aryl—aryl coupling is accomplished using a solution-based approach and investigated by scanning tunneling microscopy. Drop-casting of the TIPB monomer onto Au(111) at room temperature results in poorly ordered noncovalent arrangements of molecules and partial dehalogenation. However, drop-casting on a preheated Au(111) substrate yields various topologically distinct covalent aggregates and networks. Interestingly, some of these covalent nanostructures do not adsorb directly on the Au(111) surface, but are loosely bound to a disordered layer of a mixture of chemisorbed iodine and molecules, a conclusion that is drawn from STM data and supported by X-ray photoelectron spectroscopy. We argue that the gold surface becomes covered by a strongly chemisorbed iodine monolayer which eventually inhibits further polymerization.

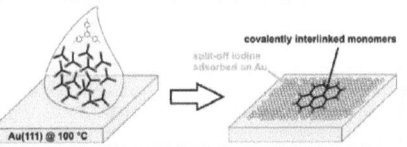

KEYWORDS: polymerization · liquid—solid · homolysis · halogenated monomer · STM · XPS · Au(111)

Over recent years the linking of small organic molecules on surfaces to form 1D and 2D polymeric structures has attracted great interest,[1−8] most lately due to the demonstration that this provides a route to nanostructured graphene with controlled dimensions.[9,10] A promising approach to generate 2D polymers is based on an Ullmann-type reaction, where halogenated monomers are covalently interlinked with the aid of a metal catalyst.[11] In this approach, the weakly bound halogen substituents are homolytically cleaved off and the resulting radicals recombine into covalent networks.[5,12−14] The network dimensionality and topology is thus determined by the halogen substitution pattern of the monomer. In the surface variant of the Ullmann reaction, the substrate serves as both the catalyst for homolysis and supporting template for the resulting network.

To date, this route is predominately pursued in ultrahigh vacuum (UHV) environments on coinage metal surfaces.[7,13,15,16] However, successful on-surface polymerization has already been demonstrated at the liquid—solid interface using alternative reaction strategies such as the Schiff base reaction[17] or boronic acid condensation,[18,19] which has been previously used for the synthesis of crystalline porous bulk materials, termed covalent organic frameworks.[20,21] Boronic acid condensation was also demonstrated on surfaces and yields well-ordered, two-dimensional covalent networks with domain sizes up to ∼50 nm, although the π-conjugation, obtainable through the Ullmann reaction, remains elusive for this approach. In addition, owing to the reversible nature of the employed condensation reactions, the resulting covalent networks only exhibit limited chemical stability.[22] Other approaches under ambient conditions encompass electrochemical epitaxial polymerization as well as light and local probe-induced polymerization.[6,23,24] Yet, these

* Address correspondence to
markus@lackinger.org,
peter.beton@nottingham.ac.uk.

Received for review July 31, 2012
and accepted March 10, 2013.

Published online March 11, 2013
10.1021/nn400337v

©2013 American Chemical Society

approaches remain restricted to specifically designed model systems. The development of a more general and flexible approach to the on-surface formation of more robust and functional 1D and 2D polymers motivates the transfer of the Ullmann-type reaction discussed above from UHV to a liquid environment. Initial attempts to achieve this goal have been limited to the formation of dimers and more extended structures have not yet been demonstrated.[25]

We have therefore studied the covalent linking of the monomer 1,3,5-tri(4-iodophenyl)benzene (TIPB) on Au(111). TIPB is a chemically stable triply iodinated organic building block that is composed of four phenyl rings and three terminating iodine atoms (the structure is shown in Figure 1a). Au(111) is chosen as a substrate because it is the best compromise between inertness against ambient contamination and sufficiently high catalytic activity for the dehalogenation reaction.[26] Furthermore, recent polymerization studies with the same compound on Au(111) under UHV conditions[27] facilitate a direct comparison between the two approaches and in addition with previous studies of 1,3,5-tri(4-bromophenyl)benzene (TBPB), the brominated analogue of TIPB.[12,14,25] As discussed above, previous studies of TBPB in a solution environment yielded only ordered arrangements of covalently interlinked dimers but very few higher oligomers and no larger covalent aggregates.[25] Consequently, in the present work we enhance the monomer reactivity by substitution of bromine with iodine, thereby taking advantage of the lower bond dissociation energy of the C—I bond and explore possibilities to form more extended covalent structures. Results of drop-cast deposition of the monomer onto the substrate held at room temperature are compared with results obtained on substrates that were preheated to 100 °C. The high resolution structural characterization by STM is augmented by chemical characterization using X-ray photoelectron spectroscopy (XPS).

RESULTS AND DISCUSSION

In Figure 1a,b we show STM images of the surface acquired following room temperature deposition of TIPB dissolved in nonanoic acid (9A) from solutions with different concentrations. These images were acquired while the sample was still covered with a liquid film with the STM tip immersed into solution. For the lower concentration (0.02 mmol/L; Figure 1a), the images show predominantly monolayer coverage with some short-range ordering. The angle between differently oriented domains is a multiple of 30° and, thus, indicates formation of nonequivalent rotational domains (see Figure 1a and Supporting Information). Higher resolution images (Figure 1a, lower middle inset) show trigonal features in a quasi-close packed arrangement. These features are highly reminiscent of that recently reported for the brominated analogue

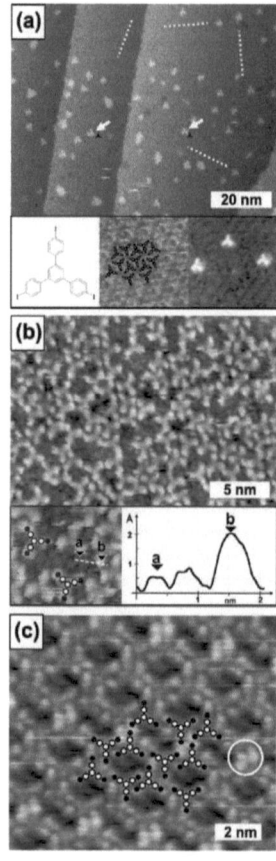

Figure 1. STM topographs of TIPB on (a, b) Au(111) and (c) graphite(0001). All images were acquired in 9A solution. (a) Obtained with a concentration of c = 0.02 mmol/L. The STM image represents small domains of ordered TIPB molecules, where individual adsorbed TIPB molecules in the second layer are highlighted by white arrows. White dotted lines indicate orientations of first layer domains. The insets depict the chemical structures of TIPB, a domain of TIPB molecules directly adsorbed on the Au(111) surface, and a zoom-in of second layer TIPB. (b) Obtained with a concentration of c = 0.04 mmol/L, resulting in an increased coverage of the second layer. The inset shows a zoom-in, where some features of the first layer are still visible. The terminating iodine atoms are highlighted by white dots. The line-profile shows a clear dip in the monolayer features, confirming the formation of a bilayer system. (c) Obtained with saturated solution on graphite. A molecular overlay indicates the tentative arrangement within the well-ordered monolayer. STM tunneling parameters: (a) I_t = 12 pA, U_t = 0.26 V; (b) I_t = 61 pA, U_t = −0.62 V; (c) I_t = 46 pA, U_t = 0.74 V.

TBPB both in UHV and in layers formed by drop-casting ethanolic solutions.[12,14,25] We attribute these features to intact TIPB monomers. Also present in Figure 1a are isolated bright features (examples are marked by white arrows) that, as shown in the lower right inset, have a clear 3-fold symmetry. These are identified as second layer TIPB molecules and the experimentally measured vertex–vertex dimensions of 1.4 ± 0.1 nm are in excellent agreement with the geometry optimized structure of intact TIPB molecules with an iodine–iodine distance of approximately 1.4 nm.

For higher concentrations (0.04 mmol/L; Figure 1b), we observe a higher density of TIPB in the second layer with an apparent height of ∼0.13 nm (see profile inset) with respect to the first monolayer. While there is no long-range ordering in the second layer, the image in Figure 1b shows features with a clear bright protrusion at each molecular lobe which are assigned to the peripheral iodine atoms of single molecules. It is noteworthy that, for self-assembly at the liquid–solid interface, stable adsorption in the second layer is rather uncommon and has only been observed in very few systems.[28]

Although highly disordered, it is interesting to compare the relative placement of molecular pairs in the second layer with those in the first layer and also in TIPB monolayers adsorbed on graphite. For this reason we show in Figure 1c a representative STM image of a self-assembled monolayer of TIPB on graphite using 9A as a solvent. This structure is well-ordered and densely packed. The molecular overlay indicates the structural model and the bright protrusions marked by the white circles are identified as cyclic arrangements of four iodine atoms that stabilize the structure by halogen–halogen interactions.[29,30] For adsorption of TIPB on the second layer on Au(111), we observe similar local arrangements where three or four iodine atoms meet. Another common junction observed in Figure 1a, two trigonal features meeting end-to-end, is also observed in the second layer; see for example the molecular pair at the bottom of the right inset to Figure 1b. In addition there are molecular junctions on the TIPB monolayer on graphite where an iodine atom from one molecule sits between two iodine atoms on a neighboring molecule (see molecular pair at the bottom center of the overlaid schematic of Figure 1c). There are many examples of this motif in Figure 1b. Overall we suggest that the second layer TIPB may be considered as a disordered version of the nanoporous monolayer physisorbed on graphite. Note also in Figure 1b the molecular arrangement forms partially completed nanopores with comparable dimensions and, in some cases, shapes to those formed on graphite. From the above discussion it is thus concluded that at room temperature the second layer TIPB molecules remain intact on the surface and the halogen substituents are not split-off. We are not able to resolve individual molecules in the underlying layer at this concentration.

In order to promote covalent interlinking, the Au(111) substrate was preheated to 100 °C on a hot plate under ambient conditions and 5 μL of TIPB solution with a notably higher concentration of c = 0.80 mmol/L was drop-cast on the surface. The sample remained on the hot plate for ∼120 s and was then allowed to cool down under ambient conditions. After this procedure, the solvent was almost fully evaporated. The sample was immediately characterized by STM and various aggregates such as 1D chains, open rings, closed pentagons, hexagons, heptagons, as well as more extended and irregular networks are clearly recognizable. Representative examples are depicted in Figure 2a. An analysis of the separation of the 3-fold vertices within this network is consistent with the formation of covalent aryl–aryl bonds. The experimental value, 1.3 ± 0.1 nm was found to be common in all types of aggregates. This value is in excellent agreement with both the figure calculated using density functional theory[27] and UHV experiments on topologically similar covalent networks,[12] verifying covalent bond formation.

The domain size observed in Figure 2a is comparable with those formed in UHV, but the total area covered by the covalent networks is smaller.[27] Moreover the formation of domains with up to 25 molecules with lateral dimensions up to 10 nm represents a major step forward from previous liquid studies, where no closed polygons were formed.

A more detailed analysis of the covalent structures reveals that several molecular lobes that do not take part in the covalent interlinks are often still terminated by iodine. Incomplete dehalogenation might be a possible reason for premature termination of the polymerization, hence mostly resulting in oligomers of finite size. Also the STM images predominantly show aggregates on substrate terraces, suggesting that the reaction is not restricted to step-edges. Changing the solvent to a shorter fatty acid, namely, heptanoic acid, while applying the same preparation protocol yielded similar results (see Supporting Information).

The catalytic role of the metallic Au(111) substrate in iodine homolysis was confirmed through control experiments with similar deposition protocols using as substrates both graphite(0001) and Au(111) intentionally terminated with iodine. The latter sample was prepared by immersing a freshly flame annealed Au(111) sample into 3 mM aqueous KI solution for 180 s and subsequent rinsing with ethanol (cf. Supporting Information). Drop-casting of TIPB solution on these substrates preheated to 100 °C did not yield any covalent structures. Both surfaces proved inactive for iodine homolysis, confirming the important catalytic contribution of the bare Au(111) surface, and,

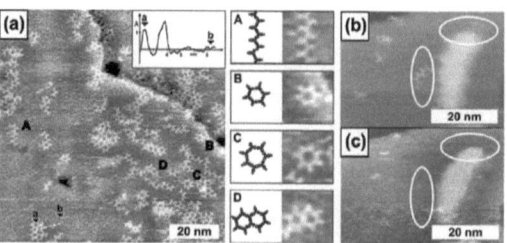

Figure 2. (a) STM topograph of covalently interlinked structures from TIPB polymerization; the sample was prepared by deposition of 5 μL of TIPB solution in 9A (c = 0.80 mmol/L) onto Au(111) held at 100 °C and cooling down after 120 s. The STM image depicts covalent aggregates on top of a first monolayer; close-ups (A−D) of frequently encountered covalent aggregates are presented in the middle: (A) one-dimensional chains, (B) closed hexagons, (C) closed heptagons, and (D) more extended structures as merged rings; besides that, more extended irregular structures and open rings were frequently observed. (b, c) Consecutively recorded STM images revealing the detachment of covalent aggregates. As highlighted by white circles covalently interlinked aggregates that were still present in (b) have disappeared in (c). STM tunneling parameters: (a) I_t = 63 pA, U_t = −0.752 V; (b, c) I_t = 58 pA, U_t = −0.624 V.

moreover, showing that Au(111) becomes catalytically inactive by iodine adsorption.

For deposition onto bare Au(111) held at 100 °C, some of the covalent aggregates are not directly adsorbed on the metal surface, but on top of an intermediate monolayer. This conclusion is supported by the STM data, where covalent aggregates appear ∼0.10−0.20 nm higher than the first monolayer. On the contrary, covalent structures obtained in UHV experiments are directly supported on metal surfaces and appear lower than the split-off iodine atoms in STM images.[27] The STM appearance of the covalent structures in Figure 2 directly compares to those obtained in polymerization studies carried out previously on iodine-terminated Au(111).[17,31,32] Furthermore, in UHV, the contrast within closed pores of covalent networks either appears significantly lower than the polymers or the halogens, that is, indicating empty pores, or with inhomogeneous features due to entrapment of molecules or atoms within the pores, whereas in Figure 2a the pore interiors appear with uniform contrast and the same height as the surrounding. In addition, the covalent aggregates are rather weakly adsorbed, as shown by the occasional observation of their detachment and displacement during STM imaging (compare Figure 2b and c).

Additional STM experiments were carried out in order to investigate the origin of this monolayer. Figure 3 shows an STM topograph acquired after the sample was annealed at 100 °C for 120 min. We interpret this image as showing adsorbed iodine on Au(111) intermingled with an array of dark pores. Higher magnification images show a more detailed view of the directly adsorbed first monolayer; see close-up in Figure 3c. Examples for frequently observed pairs of spherical features with 0.5 nm spacing are highlighted by pairs of blue arrows. This distance cannot be matched with any intramolecular distance of TIPB or its networks.

In addition, the 2D FFT of this STM image indicates hexagonal symmetry. Interestingly, there is a clearly visible hexagonally arranged group of inner spots (marked by the dashed circle in Figure 3c), but also faint outer spots can be recognized (two spots are exemplarily marked by arrows). The inner spots correspond to a period of ∼1.7 nm, indicating both a more or less regular spacing and a nonrandom azimuthal orientation of the dark pores. The outer spots correspond to a period of ∼0.5 nm and can be separated into two groups of hexagonally arranged spots that are rotated by ∼16° with respect to each other. The period is similar to the nearest neighbor iodine−iodine distance in the various coverage-dependent superstructures found for pure iodine monolayers on Au(111).[33] The presence of chemisorbed iodine within the first monolayer is further substantiated by XPS data (vide infra). Based on these experiments, we propose that iodine from cleaved bonds adsorbs on, and poisons the Au(111) surface thus inhibiting any further catalytically supported iodine homolysis. We further propose that chemisorbed iodine also displaces partially formed covalent networks from the surface leading to weak adsorption in a second layer.

XPS experiments were performed in order to confirm the presence of chemisorbed iodine on the gold surface. Different sample preparations were compared: drop-casting TIPB in 9A solutions with two different concentrations (0.02 and 0.80 mmol/L) onto Au(111) either held at room temperature or heated to 100 °C. First, 5 μL solution were applied to a freshly flame annealed Au(111), then after 2 h, where the surfaces were held at the respective temperature, the samples were rinsed with pure ethanol and immediately transferred to the XPS chamber. No further annealing or further treatment was carried out before the measurement.

Spectra highlighting the binding energy region of the I(3d) core levels are depicted in Figure 4. In the XPS

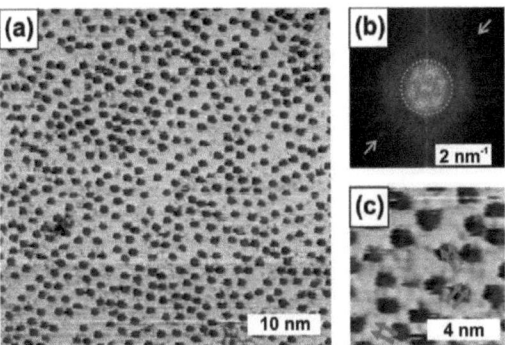

Figure 3. (a) STM topograph obtained after drop-casting 0.80 mmol/L TIPB solution in 9A onto Au(111) at a surface temperature of 100 °C and extended heating for 120 min (b) corresponding FFT to (a); the inner hexagonally arranged group of diffuse spots (marked by the dashed circle) corresponds to a real space distance of ∼1.7 nm and indicates a regular spacing and a preferred azimuthal orientation of the dark pores. The outer faint spots (examples marked by arrows) correspond to a real space distance of ∼0.5 nm. The outer spots consist of two hexagonally arranged groups that are rotated by 16° with respect to each other. (c) Close-up to (a); the pairs of blue arrows highlight 0.5 nm spaced spherical features that are assigned to nearest neighbor chemisorbed iodine atoms. STM tunneling parameters: I_t = 35 pA, U_t = −0.359 V.

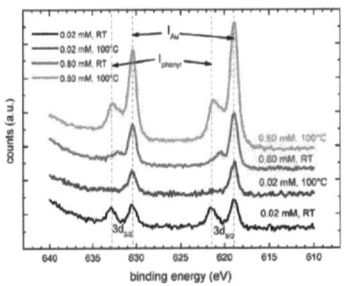

Figure 4. XPS measurements of TIPB deposited onto Au(111) from 9A solution under ambient conditions. Two different concentrations, 0.02 and 0.80 mmol/L, were applied and the Au(111) surface was either held at room temperature or preheated to 100 °C. After maintaining the respective surface temperature for 2 h, the samples were rinsed with pure ethanol. The curves are vertical offset for clarity, but were not normalized otherwise. All spectra were acquired under similar conditions, thus intensities directly correspond to iodine concentrations. The spectra show I $3d_{3/2}$ and $3d_{5/2}$ spin−orbit doublets and typical binding energies for the chemically distinct iodine species I_{phenyl} and I_{Au} are indicated by the vertical dashed lines, respectively.

data of room temperature deposited TIPB with 0.02 mmol/L (black curve in Figure 4) four peaks can be discerned that correspond to spin−orbit doublets of two chemically distinct iodine species. The spin−orbit doublet at binding energies of 621.4 eV (I-$3d_{5/2}$) and 632.8 eV (I-$3d_{3/2}$) arises from unreacted TIPB with binding energies comparable to similar iodinated aromatic compounds.[34] This iodine species is referred to as I_{phenyl} in the following. The additional spin−orbit doublet at 619.0 eV (I-$3d_{5/2}$) and 630.5 eV (I-$3d_{3/2}$) is attributed to iodine chemisorbed on Au(111) as identical binding energies were obtained in XPS control measurements on iodine terminated Au(111) surfaces (cf. Supporting Information). These binding energies are also consistent with XPS spectra for iodine adsorbed on other coinage metals, where the characteristic I-$3d_{5/2}$ peaks were found at 619.0 eV for Cu−I and 619.4 eV for Ag−I, respectively.[35] Chemisorbed iodine is referred to as I_{Au} in the following. The presence of I_{Au} for room temperature drop-cast samples indicates that spontaneous dehalogenation of TIPB molecules already occurs without additional thermal activation. However, drop-casting a low concentration solution on Au(111) held at 100 °C (blue curve in Figure 4), leads to the disappearance of I_{phenyl}. This indicates amounts of unreacted or partially reacted TIPB below the detection limit, while the I_{Au} peak increases in intensity. The onset of dehalogenation at room temperature is an obvious limitation for self-assembly of well-ordered TIBP structures on Au(111).

The XPS measurements obtained after room temperature drop-casting 0.80 mmol/L solution appear different (red curve in Figure 4). While I_{Au} is clearly present, I_{phenyl} is only visible as a small shoulder, indicating an enhanced dehalogenation rate at room temperature for higher concentrations. Interestingly, I_{Au} increases for the high concentration, 100 °C drop-cast samples (green curve in Figure 4), but also I_{phenyl} becomes slightly more prominent. This confirms larger amounts of chemisorbed iodine when solutions with higher solute concentration are applied and implies that for low concentration the Au(111) surface is not saturated with iodine. Enhancement of I_{phenyl} can be explained by the STM data, which show that for higher concentrations the covalent aggregates become larger and, accordingly, less soluble. These data provide explicit support for our interpretation of the STM image shown in Figure 3.

The onset of the reaction at room temperature is in accordance with UHV experiments on 1,3,5-triiodobenzene on Au(111), where covalently interlinked aggregates were likewise already observed at room temperature.[27] In general, the Au(111) surface is known to act as a catalyst for the homolysis of carbon—halogen bonds.[8,9,36] However, in contrast to iodine homolysis, the cleavage of bromine substituents on Au(111) requires thermal activation, as demonstrated in UHV and ambient conditions at temperatures of 140 to 180 °C[12] and 200 °C,[25] respectively. The reactivity difference between iodinated versus brominated monomers can be rationalized by pronounced differences in the carbon—halogen bond dissociation energy, with C—Br bonds (3.49 eV in bromobenzene) being significantly stronger than the C—I bonds (2.84 eV in iodobenzene).[37] Whereas, according to a systematic DFT study, the binding energy of bromine on Au(111) amounts to 2.26 eV and is even slightly higher than that of iodine of 2.14 eV.[38] In any case, halogens strongly chemisorb on Au(111), as further expressed by a high desorption temperature. Syomin et al. reported an I_2 desorption peak maximum at 450 °C for temperature-programmed desorption after C_6H_5I exposure on gold.[36]

Based on the experimental STM and XPS results, we propose the following reaction scheme. Even for room temperature deposition of TIPB molecules, initial dehalogenation spontaneously occurs, while the reaction rate becomes significantly enhanced at 100 °C and for higher concentrations. The catalytic activity of Au(111) is required for the activation of polymerization, as shown by control experiments on intentionally iodine-terminated Au(111) surfaces and graphite (0001). Following dehalogenation, radicals recombine and form new covalent aryl—aryl bonds. However, extended covalent aggregates are not observed after room temperature deposition, possibly due to the limited room temperature mobility of the radicals, and there is clear evidence that unreacted monomers remain on the surface. Drop-casting onto heated substrates results in covalent aggregates that are adsorbed on a monolayer of chemisorbed I and unreacted or partially reacted molecules. The presence of split-off iodine atoms is unambiguously confirmed by XPS experiments. Owing to the progressive adsorption of split-off iodine atoms onto the Au(111) surface the substrate becomes catalytically inactive for iodine homolysis suppressing further coupling reactions. The chemisorbed iodine layer not only poisons the substrate, but also displaces covalent aggregates into the second layer, where they are only adsorbed weakly.

CONCLUSION

It has been shown that the polymerization of the triply iodinated monomer TIPB can be initiated by deposition from solution onto a preheated Au(111) surface. In contrast to similar experiments under ambient conditions with TBPB, the brominated analogue of TIPB, more extended covalent structures that consist of up to 25 monomeric units were obtained, while the brominated monomer yielded only dimers, consistent with the enhanced reactivity of iodinated precursors for the proposed polymerization reaction. Interestingly, most of the covalent structures were found on top of a chemisorbed monolayer. The presence of covalent aggregates in a second layer is unique to the solution approach and has not been observed in UHV experiments. The observed displacement and detachment during STM imaging indicates a rather weak interaction with the underlying first monolayer acting as a buffer to Au(111). This observation is particularly interesting for the development of strategies to transfer these covalent aggregates from the catalytically active surfaces that are indispensable for their synthesis to alternative surfaces that are more promising in terms of applications.

Overall, the increased reactivity of iodinated compounds make them more promising candidates than their brominated analogues for polymerization under ambient conditions and permits the use of relatively inert surfaces such as gold. Our results indicate that, to further improve the order and size of the covalent networks, it will be necessary to reduce the effect of the iodine-induced deactivation of the catalytic surface. This raises several interesting scientific issues and there are several possible strategies, for example, methodologies for the partial removal of iodine through the introduction of other reagents and through the synthesis of modified monomers, which adsorb more strongly on the gold surface. The results presented here strongly motivate such studies.

MATERIALS AND METHODS

The STM experiments were conducted with an Agilent Technologies 4500 PicoPlus STM using a PicoScan controller. Commercially supplied (111) terminated gold films on mica (Georg Albert, Physical Vapor Deposition) were used as substrates and prepared by flame-annealing prior to the experiments. STM tips were mechanically cut from a platinum/iridium (80/20) wire. The atomic lattices of graphite(0001) and Au(111) were used for lateral calibration of the STM, and experimental distances were derived with an accuracy of <0.1 nm.

The monomer 1,3,5-tri(4-iodophenyl)benzene (TIPB, Sigma Aldrich) was used as supplied and dissolved in nonanoic acid (9A, Sigma Aldrich) and heptanoic acid (7A, Sigma Aldrich). During the deposition, the substrate was either held at room temperature or preheated to 100 °C on a hot plate under atmospheric conditions. Samples for XPS were prepared under ambient conditions, with the substrate either held at room temperature or preheated to 100 °C. The samples were kept at the respective temperature for 2 h, rinsed with pure ethanol, and then transferred into the XPS chamber. A Kratos AXIS ULTRA DLD instrument with a monochromated Al Kα X-ray source (1486.6 eV) was used and operated at 10 mA emission current and 12 kV anode potential. The lens mode used was hybrid-slot and pass energies 80 and 20 eV were used for the wide and high resolution scans, respectively. Spectra were acquired at room temperature.

Conflict of Interest: The authors declare no competing financial interest.

Acknowledgment. P.H.B. acknowledges useful discussions with Prof. Neil Champness (School of Chemistry, University of Nottingham) and funding was provided by the UK Engineering and Physical Sciences Research Council. G.E. is particularly grateful for financial support by the Hanns-Seidel-Stiftung and Kurt Fordan Förderverein für herausragende Begabung e.V. M.L. acknowledges funding by the DFG (LA1842/4) and Nanosystems-Initiative-Munich (NIM). We would like to thank Jürgen Dienstmaier for his help with the STM control experiments.

Supporting Information Available: Additional STM images of room temperature adsorption of TIPB, solvent influence, and iodine-terminated Au(111) surface. This material is available free of charge via the Internet at http://pubs.acs.org.

REFERENCES AND NOTES

1. Treier, M.; Fasel, R.; Champness, N. R.; Argent, S.; Richardson, N. V. Molecular Imaging of Polyimide Formation. *Phys. Chem. Chem. Phys.* **2009**, *11*, 1209–1214.
2. Jensen, S.; Greenwood, J.; Früchtl, H. A.; Baddeley, C. J. STM Investigation on the Formation of Oligoamides on Au{111} by Surface-Confined Reactions of Melamine with Trimesoyl Chloride. *J. Phys. Chem. C* **2011**, *115*, 8630–8636.
3. Treier, M.; Richardson, N. V.; Fasel, R. Fabrication of Surface-Supported Low-Dimensional Polyimide Networks. *J. Am. Chem. Soc.* **2008**, *130*, 14054–14055.
4. Weigelt, S.; Busse, C.; Bombis, C.; Knudsen, M. M.; Gothelf, K. V.; Laegsgaard, E.; Besenbacher, F.; Linderoth, T. R. Surface Synthesis of 2D Branched Polymer Nanostructures. *Angew. Chem., Int. Ed.* **2008**, *47*, 4406–4410.
5. Bieri, M.; Nguyen, M.-T.; Gröning, O.; Cai, J.; Treier, M.; Aït-Mansour, K.; Ruffieux, P.; Pignedoli, C. A.; Passerone, D.; Kastler, M.; Müllen, K.; Fasel, R. Two-Dimensional Polymer Formation on Surfaces: Insight into the Roles of Precursor Mobility and Reactivity. *J. Am. Chem. Soc.* **2010**, *132*, 16669–16676.
6. Miura, A.; De Feyter, S.; Abdel-Mottaleb, M. M. S.; Gesquiere, A.; Grim, P. C. M.; Moessner, G.; Sieffert, M.; Klapper, M.; Müllen, K.; De Schryver, F. C. Light- and STM-Tip-Induced Formation of One-Dimensional and Two-Dimensional Organic Nanostructures. *Langmuir* **2003**, *19*, 6474–6482.
7. Lipton-Duffin, J. A.; Ivasenko, O.; Perepichka, D. F.; Rosei, F. Synthesis of Polyphenylene Molecular Wires by Surface-Confined Polymerization. *Small* **2009**, *5*, 592–597.
8. Lafferentz, L.; Eberhardt, V.; Dri, C.; Africh, C.; Comelli, G.; Esch, F.; Hecht, S.; Grill, L. Controlling On-Surface Polymerization by Hierarchical and Substrate-Directed Growth. *Nat. Chem.* **2012**, *4*, 215–220.
9. Cai, J.; Ruffieux, P.; Jaafar, R.; Bieri, M.; Braun, T.; Blankenburg, S.; Muoth, M.; Seitsonen, A. P.; Saleh, M.; Feng, X.; Müllen, K.; Fasel, R. Atomically Precise Bottom-Up Fabrication of Graphene Nanoribbons. *Nature* **2010**, *466*, 470–473.
10. Yang, X.; Dou, X.; Rouhanipour, A.; Zhi, L.; Räder, H. J.; Müllen, K. Two-Dimensional Graphene Nanoribbons. *J. Am. Chem. Soc.* **2008**, *130*, 4216–4217.
11. Hla, S. W.; Bartels, L.; Meyer, G.; Rieder, K. H. Inducing All Steps of a Chemical Reaction with the Scanning Tunneling Microscope Tip: Towards Single Molecule Engineering. *Phys. Rev. Lett.* **2000**, *85*, 2777–2780.
12. Blunt, M. O.; Russell, J. C.; Champness, N. R.; Beton, P. H. Templating Molecular Adsorption Using a Covalent Organic Framework. *Chem. Commun.* **2010**, *46*, 7157–7159.
13. Grill, L.; Dyer, M.; Lafferentz, L.; Persson, M.; Peters, M. V.; Hecht, S. Nano-Architectures by Covalent Assembly of Molecular Building Blocks. *Nat. Nanotechnol.* **2007**, *2*, 687–691.
14. Gutzler, R.; Walch, H.; Eder, G.; Kloft, S.; Heckl, W. M.; Lackinger, M. Surface Mediated Synthesis of 2D Covalent Organic Frameworks: 1,3,5-Tris(4-bromophenyl)benzene on Graphite(001), Cu(111) and Ag(110). *Chem. Commun.* **2009**, 4456–4458.
15. Xi, M.; Bent, B. E. Mechanisms of the Ullmann Coupling Reaction in Adsorbed Monolayers. *J. Am. Chem. Soc.* **1993**, *115*, 7426–7433.
16. Lackinger, M.; Heckl, W. M. A STM Perspective on Covalent Intermolecular Coupling Reactions on Surfaces. *J. Phys. D: Appl. Phys.* **2011**, *44*, 464011.
17. Tanoue, R.; Higuchi, R.; Enoki, N.; Miyasato, Y.; Uemura, S.; Kimizuka, N.; Stieg, A. Z.; Gimzewski, J. K.; Kunitake, M. Thermodynamically Controlled Self-Assembly of Covalent Nanoarchitectures in Aqueous Solution. *ACS Nano* **2011**, *5*, 3923–3929.
18. Côté, A. P.; El-Kaderi, H. M.; Furukawa, H.; Hunt, J. R.; Yaghi, O. M. Reticular Synthesis of Microporous and Mesoporous 2D Covalent Organic Frameworks. *J. Am. Chem. Soc.* **2007**, *129*, 12914–12915.
19. Dienstmaier, J. F.; Gigler, A. M.; Goetz, A. J.; Knochel, P.; Bein, T.; Lyapin, A.; Reichlmaier, S.; Heckl, W. M.; Lackinger, M. Synthesis of Well-Ordered COF Monolayers: Surface Growth of Nanocrystalline Precursors versus Direct On-Surface Polycondensation. *ACS Nano* **2011**, *5*, 9737–9745.
20. Côté, A. P.; Benin, A. I.; Ockwig, N. W.; O'Keeffe, M.; Matzger, A. J.; Yaghi, O. M. Porous, Crystalline, Covalent Organic Frameworks. *Science* **2005**, *310*, 1166–1170.
21. Han, S. S.; Furukawa, H.; Yaghi, O. M.; Goddard, W. A. Covalent Organic Frameworks as Exceptional Hydrogen Storage Materials. *J. Am. Chem. Soc.* **2008**, *130*, 11580–11581.
22. Guan, C.-Z.; Wang, D.; Wan, L.-J. Construction and Repair of Highly Ordered 2D Covalent Networks by Chemical Equilibrium Regulation. *Chem. Commun.* **2012**, *48*, 2943–2945.
23. Xie, R. S.; Song, Y. H.; Wan, L. J.; Yuan, H. Z.; Li, P. C.; Xiao, X. P.; Liu, L.; Ye, S. H.; Lei, S. B.; Wang, L. Two-Dimensional Polymerization and Reaction at the Solid/Liquid Interface: Scanning Tunneling Microscopy Study. *Anal. Sci.* **2011**, *27*, 129–138.
24. Sakaguchi, H.; Matsumura, H.; Gong, H. Electrochemical Epitaxial Polymerization of Single-Molecular Wires. *Nat. Mater.* **2004**, *3*, 551–557.
25. Russell, J. C.; Blunt, M. O.; Garfitt, J. M.; Scurr, D. J.; Alexander, M.; Champness, N. R.; Beton, P. H. Dimerization of Tri(4-bromophenyl)benzene by Aryl–Aryl Coupling from Solution on a Gold Surface. *J. Am. Chem. Soc.* **2011**, *133*, 4220–4223.
26. Meyer, R.; Lemire, C.; Shaikhutdinov, S.; Freund, H. Surface Chemistry of Catalysis by Gold. *Gold Bull. (Geneva)* **2004**, *37*, 72–124.
27. Schlögl, S.; Heckl, W. M.; Lackinger, M. On-Surface Radical Addition of Triply Iodinated Monomers on Au(111) - The Influence of Monomer Size and Thermal Post-Processing. *Surf. Sci.* **2012**, *606*, 999–1004.
28. Blunt, M. O.; Russell, J. C.; Gimenez-Lopez, M. D.; Taleb, N.; Lin, X. L.; Schröder, M.; Champness, N. R.; Beton, P. H. Guest-Induced Growth of a Surface-Based Supramolecular Bilayer. *Nat. Chem.* **2011**, *3*, 74–78.
29. Walch, H.; Gutzler, R.; Sirtl, T.; Eder, G.; Lackinger, M. Material- and Orientation-Dependent Reactivity for Heterogeneously Catalyzed Carbon–Bromine Bond Homolysis. *J. Phys. Chem. C* **2010**, *114*, 12604–12609.
30. Gutzler, R.; Ivasenko, O.; Fu, C.; Brusso, J. L.; Rosei, F.; Perepichka, D. F. Halogen Bonds as Stabilizing Interactions

in a Chiral Self-Assembled Molecular Monolayer. *Chem. Commun.* **2011**, *47*, 9453–9455.
31. Sakaguchi, H.; Matsumura, H.; Gong, H.; Abouelwafa, A. M. Direct Visualization of the Formation of Single-Molecule Conjugated Copolymers. *Science* **2005**, *310*, 1002–1006.
32. Lapitan, L. D. S.; Tongol, B. J. V.; Yau, S. L. In Situ Scanning Tunneling Microscopy Imaging of Electropolymerized Poly(3,4-Ethylenedioxythiophene) on an Iodine-Modified Au(111) Single Crystal Electrode. *Electrochim. Acta* **2012**, *62*, 433–440.
33. Huang, L.; Zeppenfeld, P.; Horch, S.; Comsa, G. Determination of Iodine Adlayer Structures on Au(111) by Scanning Tunneling Microscopy. *J. Chem. Phys.* **1997**, *107*, 585–591.
34. Hla, S. W.; Rieder, K. H. STM Control of Chemical Reactions: Single-Molecule Synthesis. *Annu. Rev. Phys. Chem.* **2003**, *54*, 307–330.
35. *Practical Surface Analysis: Auger and X-ray Photoelectron Spectroscopy*, 2nd ed.; John Wiley & Sons: Chichester, U.K., 1990; Vol. 1.
36. Syomin, D.; Koel, B. E. Adsorption of Iodobenzene (C_6H_5I) on Au(111) Surfaces and Production of Biphenyl (C_6H_5-C_6H_5). *Surf. Sci.* **2001**, *490*, 265–273.
37. McMillen, D. F.; Golden, D. M. Hydrocarbon Bond Dissociation Energies. *Annu. Rev. Phys. Chem.* **1982**, *33*, 493–532.
38. Migani, A.; Illas, F. A Systematic Study of the Structure and Bonding of Halogens on Low-Index Transition Metal Surfaces. *J. Phys. Chem. B* **2006**, *110*, 11894–11906.

References

[1] G. E. Moore. Cramming More Components onto Integrated Circuits. *Electronics*, 38(1):114–117, 1965.

[2] J. R. Sheats and B. Smith. *Microlithography Science and Technology*. Marcel Dekker, 1 edition, 1998.

[3] P. A. Packan. Perspectives: Device Physics - Pushing the Limits. *Science*, 285(5436):2079–2081, 1999.

[4] N. J. Tao. Probing Potential-Tuned Resonant Tunneling through Redox Molecules with Scanning Tunneling Microscopy. *Physical Review Letters*, 76(21):4066–4069, 1996.

[5] G. Binnig, H. Rohrer, C. Gerber, and E. Weibel. Surface Studies by Scanning Tunneling Microscopy. *Physical Review Letters*, 49(1):57–61, 1982.

[6] G. Binnig, C. F. Quate, and C. Gerber. Atomic Force Microscope. *Physical Review Letters*, 56(9):930–933, 1986.

[7] F. J. Giessibl. High-Speed Force Sensor for Force Microscopy and Profilometry Utilizing a Quartz Tuning Fork. *Applied Physics Letters*, 73(26):3956–3958, 1998.

[8] S. Stepanow, M. Lingenfelder, A. Dmitriev, H. Spillmann, E. Delvigne, N. Lin, X. Deng, C. Cai, J. V. Barth, and K. Kern. Steering Molecular Organization and Host-Huest Interactions Using Two-Dimensional Nanoporous Coordination Systems. *Nature Materials*, 3(4):229–233, 2004.

[9] A. Saywell, J. K. Sprafke, L. J. Esdaile, A. J. Britton, A. Rienzo, H. L. Anderson, J. N. O'Shea, and P. H. Beton. Conformation and Packing of Porphyrin Polymer Chains Deposited Using Electrospray on a Gold Surface. *Angewandte Chemie International Edition*, 49(48):9136–9139, 2010.

[10] P. Amsalem, L. Giovanelli, J. M. Themlin, and T. Angot. Electronic and Vibrational Properties at the ZnPc/Ag(110) Interface. *Physical Review B*, 79(23):235426, 2009.

[11] F. Schreiber. Structure and Growth of Self-Assembling Monolayers. *Progress in Surface Science*, 65(5-8):151–257, 2000.

[12] C. Lee, Q. Li, W. Kalb, X.-Z. Liu, H. Berger, R. W. Carpick, and J. Hone. Frictional Characteristics of Atomically Thin Sheets. *Science*, 328(5974):76–80, 2010.

[13] R. W. Carpick and M. Salmeron. Scratching the Surface: Fundamental Investigations of Tribology with Atomic Force Microscopy. *Chemical Reviews*, 97(4):1163–1194, 1997.

References

[14] T. Han, J. M. Williams, and T. P. Beebe Jr. Chemical Bonds Studied with Functionalized Atomic Force Microscopy Tips. *Analytica Chimica Acta*, 307(2-3):365–376, 1995.

[15] J. Repp, G. Meyer, F. E. Olsson, and M. Persson. Controlling the Charge State of Individual Gold Adatoms. *Science*, 305(5683):493–495, 2004.

[16] J. Bansmann, S. H. Baker, C. Binns, J. A. Blackman, J. P. Bucher, J. Dorantes-Dávila, V. Dupuis, L. Favre, D. Kechrakos, A. Kleibert, K. H. Meiwes-Broer, G. M. Pastor, A. Perez, O. Toulemonde, K. N. Trohidou, J. Tuaillon, and Y. Xie. Magnetic and Structural Properties of Isolated and Assembled Clusters. *Surface Science Reports*, 56(6-7):189–275, 2005.

[17] P. Gambardella, S. Stepanow, A. Dmitriev, J. Honolka, F. M. F. de Groot, M. Lingenfelder, S. S. Gupta, D. D. Sarma, P. Bencok, S. Stanescu, S. Clair, S. Pons, N. Lin, A. P. Seitsonen, H. Brune, J. V. Barth, and K. Kern. Supramolecular Control of the Magnetic Anisotropy in Two-Dimensional High-Spin Fe Arrays at a Metal Interface. *Nat Mater*, 8(3):189–193, 2009.

[18] W. A. Hofer. Challenges and Errors: Interpreting High Resolution Images in Scanning Tunneling Microscopy. *Progress in Surface Science*, 71(5-8):147–183, 2003.

[19] L. Grill. Large Molecules on Surfaces: Deposition and Intramolecular STM Manipulation by Directional Forces. *Journal of Physics: Condensed Matter*, 22(8):084023, 2010.

[20] J. A. Stroscio and D. M. Eigler. Atomic and Molecular Manipulation with the Scanning Tunneling Microscope. *Science*, 254(5036):1319–1326, 1991.

[21] L. Grill. Functionalized Molecules Studied by STM: Motion, Switching and Reactivity. *Journal of Physics: Condensed Matter*, 20(5), 2008.

[22] K. Müllen and J. P. Rabe. Nanographenes as Active Components of Single-Molecule Electronics and How a Scanning Tunneling Microscope Puts them to Work. *Accounts of Chemical Research*, 41(4):511–520, 2008.

[23] T. Kudernac, S. B. Lei, J. A. A. W. Elemans, and S. De Feyter. Two-Dimensional Supramolecular Self-Assembly: Nanoporous Networks on Surfaces. *Chemical Society Reviews*, 38(2):402–421, 2009.

[24] C. Meier, K. Landfester, D. Künzel, T. Markert, A. Groß, and U. Ziener. Hierarchically Self-Assembled Host-Guest Network at the Solid-Liquid Interface for Single-Molecule Manipulation. *Angewandte Chemie International Edition*, 47(20):3821–3825, 2008.

[25] P. Cordier, F. Tournilhac, C. Soulie-Ziakovic, and L. Leibler. Self-Healing and Thermoreversible Rubber from Supramolecular Assembly. *Nature*, 451(7181):977–980, 2008.

[26] B. Ghosh and M. W. Urban. Self-Repairing Oxetane-Substituted Chitosan Polyurethane Networks. *Science*, 323(5920):1458–1460, 2009.

[27] J. V. Barth, G. Costantini, and K. Kern. Engineering Atomic and Molecular Nanostructures at Surfaces. *Nature*, 437(7059):671–679, 2005.

References

[28] J. V. Barth. Molecular Architectonic on Metal Surfaces. *Annual Review of Physical Chemistry*, 58:375–407, 2007.

[29] I. Felhősi, E. Kálmán, and P. Póczik. Corrosion Protection by Self-Assembly. *Russian Journal of Electrochemistry*, 38(3):230–237, 2002.

[30] P. E. Laibinis and G. M. Whitesides. Self-Assembled Monolayers of N-Alkanethiolates on Copper Are Barrier Films that Protect the Metal against Oxidation by Air. *Journal of the American Chemical Society*, 114(23):9022–9028, 1992.

[31] C. D. Dimitrakopoulos and P. R. L. Malenfant. Organic Thin Film Transistors for Large Area Electronics. *Advanced Materials*, 14(2):99–117, 2002.

[32] G. Ertl. Reactions at Surfaces: From Atoms to Complexity (Nobel Lecture). *Angewandte Chemie International Edition*, 47(19):3524–3535, 2008.

[33] R. T. Vang, J. V. Lauritsen, E. Lægsgaard, and F. Besenbacher. Scanning Tunneling Microscopy as a Tool to Study Catalytically Relevant Model Systems. *Chemical Society Reviews*, 37(10):2191–2203, 2008.

[34] A. Kühnle. Self-Assembly of Organic Molecules at Metal Surfaces. *Current Opinion in Colloid & Interface Science*, 14(2):157–168, 2009.

[35] C. Joachim, J. K. Gimzewski, and A. Aviram. Electronics Using Hybrid-Molecular and Mono-Molecular Devices. *Nature*, 408(6812):541–548, 2000.

[36] B. O'Regan and M. Gratzel. A Low-Cost, High-Efficiency Solar Cell Based on Dye-Sensitized Colloidal TiO_2 Films. *Nature*, 353(6346):737–740, 1991.

[37] T. Oyamada, H. Uchiuzou, S. Akiyama, Y. Oku, N. Shimoji, K. Matsushige, H. Sasabe, and C. Adachi. Lateral Organic Light-Emitting Diode With Field-Effect Transistor Characteristics. *Journal of Applied Physics*, 98(7):074506-7, 2005.

[38] T. Kudernac, N. Ruangsupapichat, M. Parschau, B. Maciá, N. Katsonis, S. R. Harutyunyan, K.-H. Ernst, and B. L. Feringa. Electrically Driven Directional Motion of a Four-Wheeled Molecule on a Metal Surface. *Nature*, 479(7372):208–211, 2011.

[39] A. E. Aliev, J. Oh, M. E. Kozlov, A. A. Kuznetsov, S. Fang, A. F. Fonseca, R. Ovalle, M. D. Lima, M. H. Haque, Y. N. Gartstein, M. Zhang, A. A. Zakhidov, and R. H. Baughman. Giant-Stroke, Superelastic Carbon Nanotube Aerogel Muscles. *Science*, 323(5921):1575–1578, 2009.

[40] B. Kang and G. Ceder. Battery Materials for Ultrafast Charging and Discharging. *Nature*, 458(7235):190–193, 2009.

[41] R. Gutzler. *Surface-Confined Molecular Self-Assembly*. PhD thesis, Ludwig-Maximilians-Universität, 2010.

[42] M. Li, K. Deng, Y. L. Yang, Q. D. Zeng, M. He, and C. Wang. Electronically Engineered Interface Molecular Superlattices: STM Study of Aromatic Molecules on Graphite. *Physical Review B*, 76(15):155438, 2007.

References

[43] M. Bowker. *Scanning Tunneling Microscopy in Surface Science, Nanoscience and Catalysis*. Weinheim, Wiley-VCH, 2010.

[44] G. Binnig, H. Rohrer, C. Gerber, and E. Weibel. 7x7 Reconstruction on Si(111) Resolved in Real Space. *Physical Review Letters*, 50(2):120–123, 1983.

[45] D. M. Eigler and E. K. Schweizer. Positioning Single Atoms with a Scanning Tunneling Microscope. *Nature*, 344(6266):524–526, 1990.

[46] D. M. Eigler, C. P. Lutz, and W. E. Rudge. An Atomic Switch Realized with the Scanning Tunneling Microscope. *Nature*, 352(6336):600–603, 1991.

[47] D. M. Eigler, P. S. Weiss, E. K. Schweizer, and N. D. Lang. Imaging Xe with a Low-Temperature Scanning Tunneling Microscope. *Physical Review Letters*, 66(9):1189–1192, 1991.

[48] M. F. Crommie, C. P. Lutz, and D. M. Eigler. Confinement of Electrons to Quantum Corrals on a Metal-Surface. *Science*, 262(5131):218–220, 1993.

[49] J. Wintterlin, J. Trost, S. Renisch, R. Schuster, T. Zambelli, and G. Ertl. Real-Time STM Observations of Atomic Equilibrium Fluctuations in an Adsorbate System: O/Ru(0001). *Surface Science*, 394(1-3):159–169, 1997.

[50] T. Tansel, A. Taranovskyy, and O. M. Magnussen. In Situ Video-STM Studies of Adsorbate Dynamics at Electrochemical Interfaces. *ChemPhysChem*, 11(7):1438–1445, 2010.

[51] H. Lüth. *Solid Surfaces, Interface and Thin Films*. Springer, 4 edition, 2001.

[52] C. J. Chen. *Introduction to Scanning Tunneling Microscopy*. Oxford University Press, 1993.

[53] J. Frommer. Scanning Tunneling Microscopy and Atomic Force Microscopy in Organic Chemistry. *Angewandte Chemie International Edition*, 31(10):1298–1328, 1992.

[54] P. K. Hansma and J. Tersoff. Scanning Tunneling Microscopy. *Journal of Applied Physics*, 61(2):R1–R23, 1987.

[55] J. Bardeen. Tunnelling from a Many-Particle Point of View. *Physical Review Letters*, 6(2):57–59, 1961.

[56] J. Tersoff and D. R. Hamann. Theory of the Scanning Tunneling Microscope. *Physical Review B*, 31(2):805–813, 1985.

[57] J. Tersoff and D. R. Hamann. Theory and Application for the Scanning Tunneling Microscope. *Physical Review Letters*, 50(25):1998–2001, 1983.

[58] G. Doyen, D. Drakova, V. Mujica, and M. Scheffler. Theory of the Scanning Tunneling Microscope. *Physica Status Solidi (a)*, 131(1):107–108, 1992.

[59] G. A. D. Briggs and A. J. Fisher. STM Experiment and Atomistic Modelling Hand in Hand: Individual Molecules on Semiconductor Surfaces. *Surface Science Reports*, 33(1-2):3–81, 1999.

References

[60] N. D. Lang. Vacuum Tunneling Current from an Adsorbed Atom. *Physical Review Letters*, 55(2):230–233, 1985.

[61] N. D. Lang. Theory of Single-Atom Imaging in the Scanning Tunneling Microscope. *Physical Review Letters*, 56(11):1164–1167, 1986.

[62] C. J. Chen. Theory of Scanning Tunneling Spectroscopy. *Journal of Vacuum Science & Technology A: Vacuum, Surfaces, and Films*, 6(2):319–322, 1988.

[63] C. J. Chen. Origin of Atomic Resolution on Metal-Surfaces in Scanning Tunneling Microscopy. *Physical Review Letters*, 65(4):448–451, 1990.

[64] P. Sautet and C. Joachim. Calculation of the Benzene on Rhodium STM Images. *Chemical Physics Letters*, 185(1-2):23–30, 1991.

[65] P. Sautet and C. Joachim. Are Electronic Interference Effects Important for STM Imaging of Substrates and Adsorbates? A Theoretical Analysis. *Ultramicroscopy*, 42:115–121, 1992.

[66] P. Sautet and C. Joachim. Electronic Transmission Coefficient for the Single-Impurity Problem in the Scattering-Matrix Approach. *Physical Review B*, 38(17):12238–12247, 1988.

[67] S. Datta, W. D. Tian, S. H. Hong, R. Reifenberger, J. I. Henderson, and C. P. Kubiak. Current-Voltage Characteristics of Self-Assembled Monolayers by Scanning Tunneling Microscopy. *Physical Review Letters*, 79(13):2530–2533, 1997.

[68] S. F. Alvarado, L. Rossi, P. Muller, P. F. Seidler, and W. Riess. STM-Excited Electroluminescence and Spectroscopy on Organic Materials for Display Applications. *IBM Journal of Research and Development*, 45(1):89–100, 2001.

[69] T. Müller. Scanning tunneling microscopy of physisorbed monolayers: From self-assembly to molecular devices. In Bharat Bhushan and Satoshi Kawata, editors, *Applied Scanning Probe Methods VI*, NanoScience and Technology, pages 1–30. Springer Berlin Heidelberg, 2007.

[70] M. Lackinger, T. Müller, T. G. Gopakumar, F. Müller, M. Hietschold, and G. W. Flynn. Tunneling Voltage Polarity Dependent Submolecular Contrast of Naphthalocyanine on Graphite. A STM Study of Close-Packed Monolayers under Ultrahigh-Vacuum Conditions. *Journal of Physical Chemistry B*, 108(7):2279–2284, 2004.

[71] J. Repp, P. Liljeroth, and G. Meyer. Coherent Electron-Nuclear Coupling in Oligothiophene Molecular Wires. *Nature Physics*, 6(12):975–979, 2010.

[72] P. H. Lippel, R. J. Wilson, M. D. Miller, C. Woll, and S. Chiang. High-Resolution Imaging of Copper-Phthalocyanine by Scanning-Tunneling Microscopy. *Physical Review Letters*, 62(2):171–174, 1989.

[73] P. Sautet. Images of Adsorbates with the Scanning Tunneling Microscope: Theoretical Approaches to the Contrast Mechanism. *Chemical Reviews*, 97(4):1097–1116, 1997.

REFERENCES

[74] N. Boulanger-Lewandowski and A. Rochefort. Intrusive STM Imaging. *Physical Review B*, 83(11):115430, 2011.

[75] C. L. Claypool, F. Faglioni, W. A. Goddard, H. B. Gray, N. S. Lewis, and R. A. Marcus. Source of Image Contrast in STM Images of Functionalized Alkanes on Graphite: A Systematic Functional Group Approach. *Journal of Physical Chemistry B*, 101(31):5978–5995, 1997.

[76] S. De Feyter, A. Gesquière, M. M. Abdel-Mottaleb, P. C. M. Grim, F. C. De Schryver, C. Meiners, M. Sieffert, S. Valiyaveettil, and K. Müllen. Scanning Tunneling Microscopy: A Unique Tool in the Study of Chirality, Dynamics, and Reactivity in Physisorbed Organic Monolayers. *Accounts of Chemical Research*, 33(8):520–531, 2000.

[77] K. Kobayashi. Moiré Pattern in Scanning Tunneling Microscopy of Monolayer Graphite. *Physical Review B*, 50(7):4749–4755, 1994.

[78] K. Kobayashi. Moiré Pattern in Scanning Tunneling Microscopy: Mechanism in Observation of Subsurface Nanostructures. *Physical Review B*, 53(16):11091–11099, 1996.

[79] A. T. N'Diaye, J. Coraux, T. N. Plasa, C. Busse, and T. Michely. Structure of Epitaxial Graphene on Ir(111). *New Journal of Physics*, 10:043033, 2008.

[80] A. K. Geim. Graphene: Status and Prospects. *Science*, 324(5934):1530–1534, 2009.

[81] K. S. Novoselov, A. K. Geim, S. V. Morozov, D. Jiang, Y. Zhang, S. V. Dubonos, I. V. Grigorieva, and A. A. Firsov. Electric Field Effect in Atomically Thin Carbon Films. *Science*, 306(5696):666–669, 2004.

[82] Q. H. Wang and M. C. Hersam. Room-Temperature Molecular-Resolution Characterization of Self-Assembled Organic Monolayers on Epitaxial Graphene. *Nature Chemistry*, 1(3):206–211, 2009.

[83] K. Besocke. An Easily Operable Scanning Tunneling Microscope. *Surface Science*, 181(1-2):145–153, 1987.

[84] J. Frohn, J. F. Wolf, K. Besocke, and M. Teske. Coarse Tip Distance Adjustment and Positioner for a Scanning Tunneling Microscope. *Review of Scientific Instruments*, 60(6):1200–1201, 1989.

[85] M. Wilms. New and Versatile Ultrahigh Vacuum Scanning Tunneling Microscope for Film Growth Experiments. *Review of Scientific Instruments*, 69(7):2696, 1998.

[86] F. Besenbacher, E. Lægsgaard, K. Mortensen, U. Nielsen, and I. Stensgaard. Compact, High-Stability, Thimble-Size Scanning Tunneling Microscope. *Review of Scientific Instruments*, 59(7):1035–1038, 1988.

[87] B. J. McIntyre, M. B. Salmeron, and G. A. Somorjai. A Scanning Tunneling Microscope That Operates at High-Pressures and High-Temperatures (430 K) and during Catalytic Reactions. *Catalysis Letters*, 14(3-4):263–269, 1992.

REFERENCES

[88] A. Z. Stieg, H. I. Rasool, and J. K. Gimzewski. A Flexible, Highly Stable Electrochemical Scanning Probe Microscope for Nanoscale Studies at the Solid-Liquid Interface. *Review of Scientific Instruments*, 79(10):103701, 2008.

[89] M. Marz, G. Goll, and H. v. Löhneysen. A Scanning Tunneling Microscope for a Dilution Refrigerator. *Review of Scientific Instruments*, 81(4):045102–7, 2010.

[90] H. Kambara, T. Matsui, Y. Niimi, and H. Fukuyama. Construction of a Versatile Ultralow Temperature Scanning Tunneling Microscope. *Review of Scientific Instruments*, 78(7):073703–5, 2007.

[91] G. Binnig and D. P. E. Smith. Single-Tube 3-Dimensional Scanner for Scanning Tunneling Microscopy. *Review of Scientific Instruments*, 57(8):1688–1689, 1986.

[92] G. Eder, S. Kloft, N. Martsinovich, K. Mahata, M. Schmittel, W. M. Heckl, and M. Lackinger. Incorporation Dynamics of Molecular Guests into Two-Dimensional Supramolecular Host Networks at the Liquid-Solid Interface. *Langmuir*, 27(22):13563–13571, 2011.

[93] M. J. Rost, L. Crama, P. Schakel, E. v. Tol, G. B. E. M. v. Velzen-Williams, C. F. Overgauw, H. t. Horst, H. Dekker, B. Okhuijsen, M. Seynen, A. Vijftigschild, P. Han, A. J. Katan, K. Schoots, R. Schumm, W. v. Loo, T. H. Oosterkamp, and J. W. M. Frenken. Scanning Probe Microscopes Go Video Rate and Beyond. *Review of Scientific Instruments*, 76(5):053710, 2005.

[94] M. Okano, K. Kajimura, S. Wakiyama, F. Sakai, W. Mizutani, and M. Ono. Vibration Isolation for Scanning Tunneling Microscopy. *Journal of Vacuum Science & Technology A: Vacuum, Surfaces, and Films*, 5(6):3313–3320, 1987.

[95] S. Griessl. *Zweidimensionale Architekturen organischer Adsorbate - Untersuchung mittels STM, LEED, TDS und Kraftfeldsimulationen*. PhD thesis, Technische Universität Chemnitz, 2003.

[96] H. Ibach. *Physics of Surfaces and Interfaces*. Springer, Berlin Heidelberg New York, 2006.

[97] M. B. Lee, Q. Y. Yang, and S. T. Ceyer. Dynamics of the Activated Dissociative Chemisorption of CH_4 and Implication for the Pressure Gap in Catalysis: A Molecular Beam - High Resolution Electron Energy Loss Study. *The Journal of Chemical Physics*, 87(5):2724–2741, 1987.

[98] H. Brune, J. Wintterlin, R. J. Behm, and G. Ertl. Surface Migration of "Hot" Adatoms in the Course of Dissociative Chemisorption of Oxygen on Al(111). *Physical Review Letters*, 68(5):624–626, 1992.

[99] S. Griessl, M. Lackinger, M. Edelwirth, M. Hietschold, and W. M. Heckl. Self-Assembled Two-Dimensional Molecular Host-Guest Architectures From Trimesic Acid. *Single Molecules*, 3(1):25–31, 2002.

[100] M. Lackinger, S. Griessl, W. A. Heckl, M. Hietschold, and G. W. Flynn. Self-Assembly of Trimesic Acid at the Liquid-Solid Interface - A Study of Solvent-Induced Polymorphism. *Langmuir*, 21(11):4984–4988, 2005.

References

[101] L. Kampschulte, M. Lackinger, A. K. Maier, R. S. K. Kishore, S. Griessl, M. Schmittel, and W. M. Heckl. Solvent Induced Polymorphism in Supramolecular 1,3,5-Benzenetribenzoic Acid Monolayers. *Journal of Physical Chemistry B*, 110(22):10829–10836, 2006.

[102] J. C. Russell, M. O. Blunt, J. M. Garfitt, D. J. Scurr, M. Alexander, N. R. Champness, and P. H. Beton. Dimerization of Tri(4-bromophenyl)benzene by Aryl-Aryl Coupling from Solution on a Gold Surface. *Journal of the American Chemical Society*, 133(12):4220–4223, 2011.

[103] G. Eder, S. Schlögl, K. Macknapp, W. M. Heckl, and M. Lackinger. A Combined Ion-Sputtering and Electron-Beam Annealing Device for the in vacuo Postpreparation of Scanning Probes. *Review of Scientific Instruments*, 82(3):033701, 2011.

[104] J. Garnaes, F. Kragh, K. A. Mørch, and A. R. Thölén. Transmission Electron-Microscopy of Scanning Tunneling Tips. *Journal of Vacuum Science & Technology A: Vacuum, Surfaces, and Films*, 8(1):441–444, 1990.

[105] P. Hoffrogge, H. Kopf, and R. Reichelt. Nanostructuring of Tips for Scanning Probe Microscopy by Ion Sputtering: Control of the Apex Ratio and the Tip Radius. *Journal of Applied Physics*, 90(10):5322–5327, 2001.

[106] J. P. Ibe, P. P. Bey, S. L. Brandow, R. A. Brizzolara, N. A. Burnham, D. P. DiLella, K. P. Lee, C. R. K. Marrian, and R. J. Colton. On the Electrochemical Etching of Tips for Scanning Tunneling Microscopy. *Journal of Vacuum Science & Technology A: Vacuum, Surfaces, and Films*, 8(4):3570–3575, 1990.

[107] A. D. Müller, F. Müller, M. Hietschold, F. Demming, J. Jersch, and K. Dickmann. Characterization of Electrochemically Etched Tungsten Tips for Scanning Tunneling Microscopy. *Review of Scientific Instruments*, 70(10):3970–3972, 1999.

[108] M. Fotino. Tip Sharpening by Normal and Reverse Electrochemical Etching. *Review of Scientific Instruments*, 64(1):159–167, 1993.

[109] A.-S. Lucier. *Preparation and Characterization of Tungsten Tips Suitable for Molecular Electronics Studies*. PhD thesis, McGill University, 2004.

[110] J. T. Yates. *Experimental Innovations in Surface Science: A Guide to Practical Laboratory Methods and Instruments*. AIP Press, Springer-Verlag, New York, 1998.

[111] S. Morishita and F. Okuyama. Sharpening of Monocrystalline Molybdenum Tips by Means of Inert-Gas Ion Sputtering. *Journal of Vacuum Science & Technology A: Vacuum, Surfaces, and Films*, 9(1):167–169, 1991.

[112] R. Zhang and D. G. Ivey. Preparation of Sharp Polycrystalline Tungsten Tips for Scanning Tunneling Microscopy Imaging. *Journal of Vacuum Science & Technology B*, 14(1):1–10, 1996.

[113] N. Ishida, A. Subagyo, A. Ikeuchi, and K. Sueoka. Holders for in situ Treatments of Scanning Tunneling Microscopy Tips. *Review of Scientific Instruments*, 80(9):093703, 2009.

References

[114] Z. Q. Yu, C. M. Wang, Y. Du, S. Thevuthasan, and I. Lyubinetsky. Reproducible Tip Fabrication and Cleaning for UHV STM. *Ultramicroscopy*, 108(9):873–877, 2008.

[115] E. Paparazzo, L. Moretto, S. Selci, M. Righini, and I. Farné. Effects of HF Attack on the Surface and Interface Microchemistry of W Tips for Use in the STM Microscope: A Scanning Auger Microscopy (SAM) Study. *Vacuum*, 52(4):421–426, 1999.

[116] S. Ernst, S. Wirth, M. Rams, V. Dolocan, and F. Steglich. Tip Preparation for Usage in an Ultra-Low Temperature UHV Scanning Tunneling Microscope. *Science and Technology of Advanced Materials*, 8(5):347–351, 2007.

[117] X. Li, W. Cai, J. An, S. Kim, J. Nah, D. Yang, R. Piner, A. Velamakanni, I. Jung, E. Tutuc, S. K. Banerjee, L. Colombo, and R. S. Ruoff. Large-Area Synthesis of High-Quality and Uniform Graphene Films on Copper Foils. *Science*, 324(5932):1312–1314, 2009.

[118] M. Ishigami, J. H. Chen, W. G. Cullen, M. S. Fuhrer, and E. D. Williams. Atomic Structure of Graphene on SiO2. *Nano Letters*, 7(6):1643–1648, 2007.

[119] E. C. H. Sykes. Surface Assembly: Graphene Goes Undercover. *Nature Chemistry*, 1(3):175–176, 2009.

[120] A. Reina, X. Jia, J. Ho, D. Nezich, H. Son, V. Bulovic, M. S. Dresselhaus, and J. Kong. Large Area, Few-Layer Graphene Films on Arbitrary Substrates by Chemical Vapor Deposition. *Nano Letters*, 9(1):30–35, 2008.

[121] H. Walch, A. K. Maier, W. M. Heckl, and M. Lackinger. Isotopological Supramolecular Networks from Melamine and Fatty Acids. *Journal of Physical Chemistry C*, 113(3):1014–1019, 2009.

[122] B. Venkataraman, J. J. Breen, and G. W. Flynn. Scanning-Tunneling-Microscopy Studies of Solvent Effects on the Adsorption and Mobility of TriacontaneTriacontanol Molecules Adsorbed on Graphite. *Journal of Physical Chemistry*, 99(17):6608–6619, 1995.

[123] T. Müller, G. W. Flynn, A. T. Mathauser, and A. V. Teplyakov. Temperature-Programmed Desorption Studies of N-Alkane Derivatives on Graphite: Desorption Energetics and the Influence of Functional Groups on Adsorbate Self-Assembly. *Langmuir*, 19(7):2812–2821, 2003.

[124] S. Weigelt, C. Bombis, C. Busse, M. M. Knudsen, K. V. Gothelf, E. Lægsgaard, F. Besenbacher, and T. R. Linderoth. Molecular Self-Assembly from Building Blocks Synthesized on a Surface in Ultrahigh Vacuum: Kinetic Control and Topo-Chemical Reactions. *ACS Nano*, 2(4):651–660, 2008.

[125] M. O. Blunt, J. C. Russell, M. d. C. Giménez-López, J. P. Garrahan, X. Lin, M. Schröder, N. R. Champness, and P. H. Beton. Random Tiling and Topological Defects in a Two-Dimensional Molecular Network. *Science*, 322(5904):1077–1081, 2008.

[126] S. Titmuss, A. Wander, and D. A. King. Reconstruction of Clean and Adsorbate-Covered Metal Surfaces. *Chemical Reviews*, 96(4):1291–1305, 1996.

REFERENCES

[127] N. Takeuchi, C. T. Chan, and K. M. Ho. Au(111) - A Theoretical Study of the Surface Reconstruction and the Surface Electronic Structure. *Physical Review B*, 43(17):13899–13906, 1991.

[128] J. Zhang, Y.-E. Sung, P. A. Rikvold, and A. Wieckowski. Underpotential Deposition of Cu on Au(111) in Sulfate-Containing Electrolytes: A Theoretical and Experimental Study. *The Journal of Chemical Physics*, 104(14):5699–5712, 1996.

[129] J. G. Xu and X. W. Wang. Study of Copper Underpotential Deposition on Au(111) Surfaces. *Surface Science*, 408(1-3):317–325, 1998.

[130] T. Hachiya, H. Honbo, and K. Itaya. Detailed Underpotential Deposition of Copper on Gold(III) in Aqueous Solutions. *Journal of Electroanalytical Chemistry and Interfacial Electrochemistry*, 315(1-2):275–291, 1991.

[131] G. Nagy and T. Wandlowski. Double Layer Properties of Au(111)/H_2SO_4 (Cl) + Cu^{2+} from Distance Tunneling Spectroscopy. *Langmuir*, 19(24):10271–10280, 2003.

[132] L. Huang, P. Zeppenfeld, S. Horch, and G. Comsa. Determination of Iodine Adlayer Structures on Au(111) by Scanning Tunneling Microscopy. *The Journal of Chemical Physics*, 107(2):585–591, 1997.

[133] G. E. Isted and D. S. Martin. Preparation and Characterisation of Au(110) and Cu(110) Surfaces for Applications in Ambient Environments. *Applied Surface Science*, 252(5):1883–1890, 2005.

[134] J. Lapujoulade. The Roughening of Metal Surfaces. *Surface Science Reports*, 20(4):195–249, 1994.

[135] J. C. Swarbrick, J. B. Taylor, and J. N. O'Shea. Electrospray Deposition in Vacuum. *Applied Surface Science*, 252(15):5622–5626, 2006.

[136] M. C. O'Sullivan, J. K. Sprafke, D. V. Kondratuk, C. Rinfray, T. D. W. Claridge, A. Saywell, M. O. Blunt, J. N. O'Shea, P. H. Beton, M. Malfois, and H. L. Anderson. Vernier Templating and Synthesis of a 12-Porphyrin Nano-Ring. *Nature*, 469(7328):72–75, 2011.

[137] L. Grill, I. Stass, K. H. Rieder, and F. Moresco. Preparation of Self-Ordered Molecular Layers by Pulse Injection. *Surface Science*, 600(11):143–147, 2006.

[138] S. R. Forrest. Ultrathin Organic Films Grown by Organic Molecular Beam Deposition and Related Techniques. *Chemical Reviews*, 97(6):1793–1896, 1997.

[139] C. J. Villagomez, T. Sasaki, J. M. Tour, and L. Grill. Bottom-Up Assembly of Molecular Wagons on a Surface. *Journal of the American Chemical Society*, 132(47):16848–16854, 2010.

[140] R. Gutzler, W. M. Heckl, and M. Lackinger. Combination of a Knudsen Effusion Cell with a Quartz Crystal Microbalance: In Situ Measurement of Molecular Evaporation Rates with a Fully Functional Deposition Source. *Review of Scientific Instruments*, 81(1):015108, 2010.

[141] M. Knudsen. Das Cosinusgesetz in der kinetischen Gastheorie. *Annalen Der Physik*, 353(24):1113–1121, 1916.

[142] G. Sauerbrey. Verwendung von Schwingquarzen zur Wägung dünner Schichten und zur Mikrowägung. *Zeitschrift für Physik*, 155(2):206–222, 1959.

[143] A. Freedman, P. L. Kebabian, Z. M. Li, W. A. Robinson, and J. C. Wormhoudt. Apparatus for Determination of Vapor Pressures at Ambient Temperatures Employing a Knudsen Effusion Cell and Quartz Crystal Microbalance. *Measurement Science & Technology*, 19(12):125102, 2008.

[144] M. V. Roux, M. Temprado, J. S. Chickos, and Y. Nagano. Critically Evaluated Thermochemical Properties of Polycyclic Aromatic Hydrocarbons. *Journal of Physical and Chemical Reference Data*, 37(4):1855–1996, 2008.

[145] J. M. Lehn. Supramolecular Chemistry. *Science*, 260(5115):1762–1763, 1993.

[146] G. M. Whitesides, J. P. Mathias, and C. T. Seto. Molecular Self-Assembly and Nanochemistry - A Chemical Strategy for the Synthesis of Nanostructures. *Science*, 254(5036):1312–1319, 1991.

[147] L. Pirondini, A. G. Stendardo, S. Geremia, M. Campagnolo, P. Samori, J. P. Rabe, R. Fokkens, and E. Dalcanale. Dynamic Materials Through Metal-Directed and Solvent-Driven Self-Assembly of Cavitands. *Angewandte Chemie International Edition*, 42(12):1384–1387, 2003.

[148] L. Piot, A. Marchenko, J. S. Wu, K. Müllen, and D. Fichou. Structural Evolution of Hexa-Peri-Hexabenzocoronene - Adlayers in Heteroepitaxy on N-Pentacontane Template Monolayers. *Journal of the American Chemical Society*, 127(46):16245–16250, 2005.

[149] J. Schnadt, W. Xu, R. Vang, J. Knudsen, Z. Li, E. Lægsgaard, and F. Besenbacher. Interplay of Adsorbate-Adsorbate and Adsorbate-Substrate Interactions in Self-Assembled Molecular Surface Nanostructures. *Nano Research*, 3(7):459–471, 2010.

[150] R. Otero, F. Hummelink, F. Sato, S. B. Legoas, P. Thostrup, E. Lægsgaard, I. Stensgaard, D. S. Galvao, and F. Besenbacher. Lock-and-Key Effect in the Surface Diffusion of Large Organic Molecules Probed by STM. *Nature Materials*, 3(11):779–782, 2004.

[151] M. Schunack, T. R. Linderoth, F. Rosei, E. Lægsgaard, I. Stensgaard, and F. Besenbacher. Long Jumps in the Surface Diffusion of Large Molecules. *Physical Review Letters*, 88(15):156102, 2002.

[152] F. Ullmann and J. Bielecki. Ueber Synthesen in der Biphenylreihe. *Berichte der deutschen chemischen Gesellschaft*, 34(2):2174–2185, 1901.

[153] F. Ullmann. Ueber eine neue Bildungsweise von Diphenylaminderivaten. *Berichte der deutschen chemischen Gesellschaft*, 36(2):2382–2384, 1903.

[154] Y. Wan, J. Chen, D. Zhang, and H. Li. Ullmann Coupling Reaction in Aqueous Conditions over the Ph-MCM-41 Supported Pd Catalyst. *Journal of Molecular Catalysis A: Chemical*, 258(1-2):89–94, 2006.

REFERENCES

[155] M. M. Blake, S. U. Nanayakkara, S. A. Claridge, L. C. Fernandez-Torres, E. C. H. Sykes, and P. S. Weiss. Identifying Reactive Intermediates in the Ullmann Coupling Reaction by Scanning Tunneling Microscopy and Spectroscopy. *Journal of Physical Chemistry A*, 113(47):13167–13172, 2009.

[156] M. Xi and B. E. Bent. Mechanisms of the Ullmann Coupling Reaction in Adsorbed Monolayers. *Journal of the American Chemical Society*, 115(16):7426–7433, 1993.

[157] M. Xi and B. E. Bent. Iodobenzene on Cu(111): Formation and Coupling of Adsorbed Phenyl Groups. *Surface Science*, 278(1-2):19–32, 1992.

[158] E. Sperotto, G. P. M. van Klink, G. van Koten, and J. G. de Vries. The Mechanism of the Modified Ullmann Reaction. *Dalton Transactions*, 39(43), 2010.

[159] M. Luo, W. Lu, D. Kim, E. Chu, J. Wyrick, C. Holzke, D. Salib, K. D. Cohen, Z. Cheng, D. Sun, Y. Zhu, T. L. Einstein, and L. Bartels. Coalescence of 3-Phenyl-Propynenitrile on Cu(111) into Interlocking Pinwheel Chains. *The Journal of Chemical Physics*, 135(13):134705–5, 2011.

[160] S. B. Lei, K. Tahara, F. C. De Schryver, M. Van der Auweraer, Y. Tobe, and S. De Feyter. One Building Block, Two Different Supramolecular Surface-Confined Patterns: Concentration in Control at the Solid-Liquid Interface. *Angewandte Chemie International Edition*, 47(16):2964–2968, 2008.

[161] M. Stöhr, M. Wahl, C. H. Galka, T. Riehm, T. A. Jung, and L. H. Gade. Controlling Molecular Assembly in Two Dimensions: The Concentration Dependence of Thermally Induced 2D Aggregation of Molecules on a Metal Surface. *Angewandte Chemie International Edition*, 44(45):7394–7398, 2005.

[162] Y. Ye, W. Sun, Y. Wang, X. Shao, X. Xu, F. Cheng, J. Li, and K. Wu. A Unified Model: Self-Assembly of Trimesic Acid on Gold. *The Journal of Physical Chemistry C*, 111(28):10138–10141, 2007.

[163] S. C. Jensen, A. E. Baber, H. L. Tierney, and E. C. H. Sykes. Dimethyl Sulfide on Cu(111): Molecular Self-Assembly and Submolecular Resolution Imaging. *ACS Nano*, 1(5):423–428, 2007.

[164] R. Gutzler, T. Sirtl, J. F. Dienstmaier, K. Mahata, W. M. Heckl, M. Schmittel, and M. Lackinger. Reversible Phase Transitions in Self-Assembled Monolayers at the Liquid-Solid Interface: Temperature-Controlled Opening and Closing of Nanopores. *Journal of the American Chemical Society*, 132(14):5084–5090, 2010.

[165] C. A. Hunter and J. K. M. Sanders. The Nature of $\pi - \pi$ Interactions. *Journal of the American Chemical Society*, 112(14):5525–5534, 1990.

[166] J. Björk, F. Hanke, C.-A. Palma, P. Samori, M. Cecchini, and M. Persson. Adsorption of Aromatic and Anti-Aromatic Systems on Graphene through $\pi - \pi$ Stacking. *The Journal of Physical Chemistry Letters*, 1(23):3407–3412, 2010.

[167] H. Margenau. Van der Waals Forces. *Reviews of Modern Physics*, 11(1):1–35, 1939.

[168] K. Autumn, Y. A. Liang, S. T. Hsieh, W. Zesch, W. P. Chan, T. W. Kenny, R. Fearing, and R. J. Full. Adhesive Force of a Single Gecko Foot-Hair. *Nature*, 405(6787):681–685, 2000.

[169] M. W. Cole, D. Velegol, H.-Y. Kim, and A. A. Lucas. Nanoscale van der Waals Interactions. *Molecular Simulation*, 35(10-11):849–866, 2009.

[170] F. London. Zur Theorie und Systematik der Molekularkräfte. *Zeitschrift für Physik A Hadrons and Nuclei*, 63(3):245–279, 1930.

[171] R. Eisenschitz and F. London. Über das Verhältnis der van der Waalsschen Kräfte zu den homöopolaren Bindungskräften. *Zeitschrift für Physik A Hadrons and Nuclei*, 60(7):491–527, 1930.

[172] J. E. Jones. On the Determination of Molecular Fields. II. From the Equation of State of a Gas. *Proceedings of the Royal Society of London. Series A*, 106(738):463–477, 1924.

[173] V. Nicolosi, P. D. Nellist, S. Sanvito, E. C. Cosgriff, S. Krishnamurthy, W. J. Blau, M. L. H. Green, D. Vengust, D. Dvorsek, D. Mihailovic, G. Compagnini, J. Sloan, V. Stolojan, J. D. Carey, S. J. Pennycook, and J. N. Coleman. Observation of Van der Waals Driven Self-Assembly of MoSI Nanowires into a Low-Symmetry Structure Using Aberration-Corrected Electron Microscopy. *Advanced Materials*, 19(4):543–547, 2007.

[174] P. N. Dickerson, A. M. Hibberd, N. Oncel, and S. L. Bernasek. Hydrogen-Bonding versus Van der Waals Interactions in Self-Assembled Monolayers of Substituted Isophthalic Acids. *Langmuir*, 26(23):18155–18161, 2010.

[175] K. S. Mali, K. Lava, K. Binnemans, and S. De Feyter. Hydrogen Bonding versus van der Waals Interactions: Competitive Influence of Noncovalent Interactions on 2D Self-Assembly at the Liquid-Solid Interface. *Chemistry - A European Journal*, 16(48):14447–14458, 2010.

[176] G. R. Desiraju. Hydrogen Bridges in Crystal Engineering: Interactions without Borders. *Accounts of Chemical Research*, 35(7):565–573, 2002.

[177] T. Steiner. The Hydrogen Bond in the Solid State. *Angewandte Chemie International Edition*, 41(1):48–76, 2002.

[178] J. A. K. Howard, V. J. Hoy, D. O'Hagan, and G. T. Smith. How Good Is Fluorine as a Hydrogen Bond Acceptor? *Tetrahedron*, 52(38):12613–12622, 1996.

[179] S. Gronert. Theoretical Studies of Proton Transfers. 1. The Potential Energy Surfaces of the Identity Reactions of the First- and Second-Row Non-Metal Hydrides with their Conjugate Bases. *Journal of the American Chemical Society*, 115(22):10258–10266, 1993.

[180] H. Walch. *Molecular Networks Through Surface-Mediated Reactions - From Hydrogen Bonds to Covalent Links*. PhD thesis, Ludwig-Maximilians-Universität, 2011.

[181] G. R. Desiraju. The C-H $\cdots$ O Hydrogen Bond: Structural Implications and Supramolecular Design. *Accounts of Chemical Research*, 29(9):441–449, 1996.

REFERENCES

[182] P. Gilli, V. Bertolasi, V. Ferretti, and G. Gilli. Evidence for Resonance-Assisted Hydrogen Bonding. 4. Covalent Nature of the Strong Homonuclear Hydrogen Bond. Study of the O–H–O System by Crystal Structure Correlation Methods. *Journal of the American Chemical Society*, 116(3):909–915, 1994.

[183] P. Gilli, V. Bertolasi, L. Pretto, V. Ferretti, and G. Gilli. Covalent versus Electrostatic Nature of the Strong Hydrogen Bond: Discrimination among Single, Double, and Asymmetric Single-Well Hydrogen Bonds by Variable-Temperature X-Ray Crystallographic Methods in ß-Diketone Enol RAHB Systems. *Journal of the American Chemical Society*, 126(12):3845–3855, 2004.

[184] M. Lackinger, S. Griessl, T. Markert, F. Jamitzky, and W. M. Heckl. Self-Assembly of Benzene-Dicarboxylic Acid Isomers at the Liquid Solid Interface: Steric Aspects of Hydrogen Bonding. *Journal of Physical Chemistry B*, 108(36):13652–13655, 2004.

[185] L. Kampschulte, S. Griessl, W. M. Heckl, and M. Lackinger. Mediated Coadsorption at the Liquid-Solid Interface: Stabilization through Hydrogen Bonds. *Journal of Physical Chemistry B*, 109(29):14074–14078, 2005.

[186] S. De Feyter and F. C. De Schryver. Self-Assembly at the Liquid/Solid Interface: STM Reveals. *Journal of Physical Chemistry B*, 109(10):4290–4302, 2005.

[187] A. G. Slater, P. H. Beton, and N. R. Champness. Two-Dimensional Supramolecular Chemistry on Surfaces. *Chemical Science*, 2(8):1440–1448, 2011.

[188] J. M. MacLeod, O. Ivasenko, C. Fu, T. Taerum, F. Rosei, and D. F. Perepichka. Supramolecular Ordering in Oligothiophene-Fullerene Monolayers. *Journal of the American Chemical Society*, 131(46):16844–16850, 2009.

[189] C. Heininger, L. Kampschulte, W. M. Heckl, and M. Lackinger. Distinct Differences in Self-Assembly of Aromatic Linear Dicarboxylic Acids. *Langmuir*, 25(2):968–972, 2008.

[190] P. Metrangolo and G. Resnati. Halogen Bonding: A Paradigm in Supramolecular Chemistry. *Chemistry - A European Journal*, 7(12):2511–2519, 2001.

[191] H. L. Nguyen, P. N. Horton, M. B. Hursthouse, A. C. Legon, and D. W. Bruce. Halogen Bonding: A New Interaction for Liquid Crystal Formation. *Journal of the American Chemical Society*, 126(1):16–17, 2003.

[192] P. Metrangolo, F. Meyer, T. Pilati, G. Resnati, and G. Terraneo. Halogen Bonding in Supramolecular Chemistry. *Angewandte Chemie International Edition*, 47(33):6114–6127, 2008.

[193] F. C. Pigge, V. R. Vangala, P. P. Kapadia, D. C. Swenson, and N. P. Rath. Hexagonal Crystalline Inclusion Complexes of 4-Iodophenoxy Trimesoate. *Chemical Communications*, (39):4726–4728, 2008.

[194] E. I. Howard, R. Sanishvili, R. E. Cachau, A. Mitschler, B. Chevrier, P. Barth, V. Lamour, M. Van Zandt, E. Sibley, C. Bon, D. Moras, T. R. Schneider, A. Joachimiak, and A. Podjarny. Ultrahigh Resolution Drug Design I: Details of Interactions in Human

REFERENCES

Aldose Reductase - Inhibitor Complex at 0.66 Å. *Proteins: Structure, Function, and Bioinformatics*, 55(4):792–804, 2004.

[195] P. Politzer, P. Lane, M. Concha, Y. Ma, and J. Murray. An Overview of Halogen Bonding. *Journal of Molecular Modeling*, 13(2):305–311, 2007.

[196] E. Bosch and C. L. Barnes. Triangular Halogen-Halogen-Halogen Interactions as a Cohesive Force in the Structures of Trihalomesitylenes. *Crystal Growth & Design*, 2(4):299–302, 2002.

[197] F. F. Awwadi, R. D. Willett, S. F. Haddad, and B. Twamley. The Electrostatic Nature of Aryl-Bromine-Halide Synthons: The Role of Aryl-Bromine-Halide Synthons in the Crystal Structures of the Trans-Bis(2-Bromopyridine)Dihalocopper(II) and Trans-Bis(3-Bromopyridine)Dihalocopper(II) Complexes. *Crystal Growth & Design*, 6(8):1833–1838, 2006.

[198] D. Maspoch, D. Ruiz-Molina, and J. Veciana. Old Materials with New Tricks: Multifunctional Open-Framework Materials. *Chemical Society Reviews*, 36(5):770–818, 2007.

[199] J. L. C. Rowsell and O. M. Yaghi. Metal-Organic Frameworks: A New Class of Porous Materials. *Microporous and Mesoporous Materials*, 73(1-2):3–14, 2004.

[200] D. Farrusseng, S. Aguado, and C. Pinel. Metal-Organic Frameworks: Opportunities for Catalysis. *Angewandte Chemie International Edition*, 48(41):7502–7513, 2009.

[201] N. Lin, A. Dmitriev, J. Weckesser, J. V. Barth, and K. Kern. Real-Time Single-Molecule Imaging of the Formation and Dynamics of Coordination Compounds. *Angewandte Chemie International Edition*, 41(24):4779–4783, 2002.

[202] D. B. Dougherty, P. Maksymovych, and J. T. Yates. Direct STM Evidence for Cu-Benzoate Surface Complexes on Cu(110). *Surface Science*, 600(19):4484–4491, 2006.

[203] D. Zacher, R. Schmid, C. Wöll, and R. A. Fischer. Surface Chemistry of Metal-Organic Frameworks at the Liquid-Solid Interface. *Angewandte Chemie International Edition*, 50(1):176–199, 2011.

[204] U. Schlickum, R. Decker, F. Klappenberger, G. Zoppellaro, S. Klyatskaya, M. Ruben, I. Silanes, A. Arnau, K. Kern, H. Brune, and J. V. Barth. Metal-Organic Honeycomb Nanomeshes with Tunable Cavity Size. *Nano Letters*, 7(12):3813–3817, 2007.

[205] M. Lackinger and W. M. Heckl. A STM Perspective on Covalent Intermolecular Coupling Reactions on Surfaces. *Journal of Physics D: Applied Physics*, 44(46):464011, 2011.

[206] R. S. Xie, Y. H. Song, L. L. Wan, H. Z. Yuan, P. C. Li, X. P. Xiao, L. Liu, S. H. Ye, S. B. Lei, and L. Wang. Two-Dimensional Polymerization and Reaction at the Solid/Liquid Interface: Scanning Tunneling Microscopy Study. *Analytical Sciences*, 27(2):129–138, 2011.

[207] L. Grill, M. Dyer, L. Lafferentz, M. Persson, M. V. Peters, and S. Hecht. Nano-Architectures by Covalent Assembly of Molecular Building Blocks. *Nature Nanotechnology*, 2(11):687–691, 2007.

References

[208] M. Schunack. *Scanning Tunneling Microscopy Studies of Organic Molecules on Metal Surfaces*. PhD thesis, University of Aarhus, 2002.

[209] K. A. R. Mitchell. On the Bond Lengths Reported for Chemisorption on Metal Surfaces. *Surface Science*, 100(1):225–240, 1980.

[210] D. Kreikemeyer-Lorenzo, W. Unterberger, D. A. Duncan, M. K. Bradley, T. J. Lerotholi, J. Robinson, and D. P. Woodruff. Face-Dependent Bond Lengths in Molecular Chemisorption: The Formate Species on Cu(111) and Cu(110). *Physical Review Letters*, 107(4):046102, 2011.

[211] J. Hagen. *Industrial Catalysis: A Practical Approach*. Wiley, 2006.

[212] D. M. Newns. Self-Consistent Model of Hydrogen Chemisorption. *Physical Review*, 178(3):1123–1135, 1969.

[213] P. W. Anderson. Localized Magnetic States in Metals. *Physical Review*, 124(1):41–53, 1961.

[214] B. Hammer and J. K. Nørskov. Theory of adsorption and surface reactions. In R.M. Lambert and G. Pacchioni, editors, *Chemisorption and Reactivity on Supported Clusters and Thin Films: Towards an Understanding of Microscopic Processes in Catalysis*, pages 285–351. Kluwer Academic Publishers, 1997.

[215] B. Hammer and J. K. Nørskov. Theoretical Surface Science and Catalysis - Calculations and Concepts. *Advances in Catalysis*, 45:71–129, 2000.

[216] B. Hammer and J. K. Nørskov. Why Gold Is the Noblest of All the Metals. *Nature*, 376(6537):238–240, 1995.

[217] G. A. Somorjai, K. R. McCrea, and J. Zhu. Active Sites in Heterogeneous Catalysis: Development of Molecular Concepts and Future Challenges. *Topics in Catalysis*, 18(3-4):157–166, 2002.

[218] P. Gambardella, Z. Sljivancanin, B. Hammer, M. Blanc, K. Kuhnke, and K. Kern. Oxygen Dissociation at Pt Steps. *Physical Review Letters*, 87(5):056103, 2001.

[219] Z. P. Liu and P. Hu. General Rules for Predicting Where a Catalytic Reaction Should Occur on Metal Surfaces: A Density Functional Theory Study of C-H and C-O Bond Breaking/Making on Flat, Stepped, and Kinked Metal Surfaces. *Journal of the American Chemical Society*, 125(7):1958–1967, 2003.

[220] K. Tahara, S. Furukawa, H. Uji-I, T. Uchino, T. Ichikawa, J. Zhang, W. Mamdouh, M. Sonoda, F. C. De Schryver, S. De Feyter, and Y. Tobe. Two-Dimensional Porous Molecular Networks of Dehydrobenzo[12]Annulene Derivatives via Alkyl Chain Interdigitation. *Journal of the American Chemical Society*, 128(51):16613–16625, 2006.

[221] J. L. Atwood, L. J. Barbour, and A. Jerga. Storage of Methane and Freon by Interstitial Van der Waals Confinement. *Science*, 296(5577):2367–2369, 2002.

… REFERENCES

[222] N. A. A. Zwaneveld, R. Pawlak, M. Abel, D. Catalin, D. Gigmes, D. Bertin, and L. Porte. Organized Formation of 2D Extended Covalent Organic Frameworks at Surfaces. *Journal of the American Chemical Society*, 130(21):6678–6679, 2008.

[223] J. F. Dienstmaier, A. M. Gigler, A. J. Goetz, P. Knochel, T. Bein, A. Lyapin, S. Reichlmaier, W. M. Heckl, and M. Lackinger. Synthesis of Well-Ordered COF Monolayers: Surface Growth of Nanocrystalline Precursors versus Direct On-Surface Polycondensation. *ACS Nano*, 5(12):9737–9745, 2011.

[224] S. J. H. Griessl, M. Lackinger, F. Jamitzky, T. Markert, M. Hietschold, and W. M. Heckl. Incorporation and Manipulation of Coronene in an Organic Template Structure. *Langmuir*, 20(21):9403–9407, 2004.

[225] S. Lei, M. Surin, K. Tahara, J. Adisoejoso, R. Lazzaroni, Y. Tobe, and S. D. Feyter. Programmable Hierarchical Three-Component 2D Assembly at a Liquid-Solid Interface: Recognition, Selection, and Transformation. *Nano Letters*, 8(8):2541–2546, 2008.

[226] E. Weber, M. Hecker, E. Koepp, W. Orlia, M. Czugler, and I. Csoregh. New Trigonal Lattice Hosts - Stoicheiometric Crystal Inclusions of Laterally Trisubstituted Benzenes - X-Ray Crystal-Structure of 1,3,5-Tris-(4-Carboxyphenyl)Benzene.Dimethylformamide. *Journal of the Chemical Society - Perkin Transactions 2*, (7):1251–1257, 1988.

[227] R. Gutzler, O. Ivasenko, C. Fu, J. L. Brusso, F. Rosei, and D. F. Perepichka. Halogen Bonds as Stabilizing Interactions in a Chiral Self-Assembled Molecular Monolayer. *Chemical Communications*, 47(33):9453–9455, 2011.

[228] A. C. Legon. The Halogen Bond: An Interim Perspective. *Physical Chemistry Chemical Physics*, 12(28):7736–7747, 2010.

[229] J. K. Yoon, W.-j. Son, K.-H. Chung, H. Kim, S. Han, and S.-J. Kahng. Visualizing Halogen Bonds in Planar Supramolecular Systems. *The Journal of Physical Chemistry C*, 115(5):2297–2301, 2011.

[230] D. Syomin and B. E. Koel. Adsorption of Iodobenzene (C_6H_5I) on Au(111) Surfaces and Production of Biphenyl (C_6H_5-C_6H_5). *Surface Science*, 490(3):265–273, 2001.

[231] L. Lafferentz, V. Eberhardt, C. Dri, C. Africh, G. Comelli, F. Esch, S. Hecht, and L. Grill. Controlling On-Surface Polymerization by Hierarchical and Substrate-Directed Growth. *Nature Chemistry*, 4(3):215–220, 2012.

[232] H. Walch, R. Gutzler, T. Sirtl, G. Eder, and M. Lackinger. Material- and Orientation-Dependent Reactivity for Heterogeneously Catalyzed Carbon-Bromine Bond Homolysis. *Journal of Physical Chemistry C*, 114(29):12604–12609, 2010.

[233] R. Gutzler, H. Walch, G. Eder, S. Kloft, W. M. Heckl, and M. Lackinger. Surface Mediated Synthesis of 2D Covalent Organic Frameworks: 1,3,5-Tris(4-Bromophenyl)Benzene on Graphite(001), Cu(111), and Ag(110). *Chemical Communications*, (29):4456–4458, 2009.

[234] H. S. Lee, S. Iyengar, and I. H. Musselman. Identification of Halogen Atoms in Scanning Tunneling Microscopy Images of Substituted Phenyl Octadecyl Ethers. *Analytical Chemistry*, 73(22):5532–5538, 2001.

References

[235] L. Kampschulte, T. L. Werblowsky, R. S. K. Kishore, M. Schmittel, W. M. Heckl, and M. Lackinger. Thermodynamical Equilibrium of Binary Supramolecular Networks at the Liquid-Solid Interface. *Journal of the American Chemical Society*, 130(26):8502–8507, 2008.

[236] C. N. R. Rao, A. K. Sood, R. Voggu, and K. S. Subrahmanyam. Some Novel Attributes of Graphene. *The Journal of Physical Chemistry Letters*, 1(2):572–580, 2010.

[237] A. Stannard, J. C. Russell, M. O. Blunt, C. Salesiotis, M. d. C. Giménez-López, N. Taleb, M. Schröder, N. R. Champness, J. P. Garrahan, and P. H. Beton. Broken Symmetry and the Variation of Critical Properties in the Phase Behaviour of Supramolecular Rhombus Tilings. *Nature Chemistry*, 4(2):112–117, 2012.

[238] H. Walch, J. Dienstmaier, G. Eder, R. Gutzler, S. Schlögl, T. Sirtl, K. Das, M. Schmittel, and M. Lackinger. Extended Two-Dimensional Metal-Organic Frameworks Based on Thiolate-Copper Coordination Bonds. *Journal of the American Chemical Society*, pages 7909–7915, 2011.

[239] S. Stepanow, N. Lin, and J. V. Barth. Modular Assembly of Low-Dimensional Coordination Architectures on Metal Surfaces. *Journal of Physics: Condensed Matter*, 20(18):184002, 2008.

[240] J. V. Barth, J. Weckesser, N. Lin, A. Dmitriev, and K. Kern. Supramolecular Architectures and Nanostructures at Metal Surfaces. *Applied Physics A: Materials Science & Processing*, 76(5):645–652, 2003.

[241] C. C. Perry, S. Haq, B. G. Frederick, and N. V. Richardson. Face Specificity and the Role of Metal Adatoms in Molecular Reorientation at Surfaces. *Surface Science*, 409(3):512–520, 1998.

[242] S. M. Driver and D. P. Woodruff. Scanning Tunnelling Microscopy Study of the Interaction of Dimethyl Disulphide with Cu(111). *Surface Science*, 457(1-2):11–23, 2000.

[243] A. Ferral, P. Paredes-Olivera, V. A. Macagno, and E. M. Patrito. Chemisorption and Physisorption of Alkanethiols on Cu(111). A Quantum Mechanical Investigation. *Surface Science*, 525(1-3):85–99, 2003.

[244] M. Konopka, R. Turansky, M. Dubecky, D. Marx, and I. Stich. Molecular Mechanochemistry Understood at the Nanoscale: Thiolate Interfaces and Junctions with Copper Surfaces and Clusters. *Journal of Physical Chemistry C*, 113(20):8878–8887, 2009.

[245] G. Pawin, K. L. Wong, D. Kim, D. Z. Sun, D. Bartels, S. Hong, T. S. Rahman, R. Carp, and M. Marsella. A Surface Coordination Network Based on Substrate-Derived Metal Adatoms with Local Charge Excess. *Angewandte Chemie International Edition*, 47(44):8442–8445, 2008.

[246] M. Matena, M. Stöhr, T. Riehm, J. Björk, S. Martens, M. S. Dyer, M. Persson, J. Lobo-Checa, K. Müller, M. Enache, H. Wadepohl, J. Zegenhagen, T. A. Jung, and L. H. Gade. Aggregation and Contingent Metal/Surface Reactivity of 1,3,8,10-Tetraazaperopyrene (TAPP) on Cu(111). *Chemistry - A European Journal*, 16(7):2079–2091, 2010.

[247] N. Lin, D. Payer, A. Dmitriev, T. Strunskus, C. Woll, J. V. Barth, and K. Kern. Two-Dimensional Adatom Gas Bestowing Dynamic Heterogeneity on Surfaces. *Angewandte Chemie International Edition*, 44(10):1488–1491, 2005.

[248] J. Sakamoto, J. van Heijst, O. Lukin, and A. D. Schlüter. Two-Dimensional Polymers: Just a Dream of Synthetic Chemists? *Angewandte Chemie International Edition*, 48(6):1030–1069, 2009.

[249] J. A. Lipton-Duffin, O. Ivasenko, D. F. Perepichka, and F. Rosei. Synthesis of Polyphenylene Molecular Wires by Surface-Confined Polymerization. *Small*, 5(5):592–597, 2009.

[250] R. J. Kominar, M. J. Krech, and S. J. W. Price. Pyrolysis of Bromobenzene by Toluene Carrier Technique and Determination of $D(C_6H_5 - Br)$. *Canadian Journal of Chemistry*, 56(11):1589–1592, 1978.

[251] M. Szwarc. Energy of the Central C-C Bond in Diphenyl. *Nature*, 161(4101):890–891, 1948.

[252] S. V. Ley and A. W. Thomas. Modern Synthetic Methods for Copper-Mediated C(aryl)-O, C(aryl)-N, and C(aryl)-S Bond Formation. *Angewandte Chemie International Edition*, 42(44):5400–5449, 2003.

[253] G. S. McCarty and P. S. Weiss. Formation and Manipulation of Protopolymer Chains. *Journal of the American Chemical Society*, 126(51):16772–16776, 2004.

[254] M. Giesen. Step and Island Dynamics at Solid/Vacuum and Solid/Liquid Interfaces. *Progress in Surface Science*, 68(1-3):1–153, 2001.

[255] W. W. Pai, N. C. Bartelt, M. R. Peng, and J. E. Reuttrobey. Steps as Adatom Sources for Surface-Chemistry - Oxygen Overlayer Formation on Ag(110). *Surface Science*, 330(3):679–685, 1995.

[256] C. H. Christensen and J. K. Nørskov. A Molecular View of Heterogeneous Catalysis. *Journal of Chemical Physics*, 128(18):182503, 2008.

[257] B. Hammer. Special Sites at Noble and Late Transition Metal Catalysts. *Topics in Catalysis*, 37(1):3–16, 2006.

[258] G. Eder, E. F. Smith, Wolfgang M. Heckl, P. H. Beton, and M. Lackinger. Solution Preparation of Two Dimensional Covalently Linked Networks by Polymerization of 1,3,5-Tri(4-iodophenyl)benzene on Au(111). 2012.

[259] S. W. Hla, L. Bartels, G. Meyer, and K. H. Rieder. Inducing all Steps of a Chemical Reaction with the Scanning Tunneling Microscope Tip: Towards Single Molecule Engineering. *Physical Review Letters*, 85(13):2777–2780, 2000.

[260] R. Meyer, C. Lemire, S. Shaikhutdinov, and H. Freund. Surface Chemistry of Catalysis by Gold. *Gold Bulletin (Geneva)*, 37(1):72–124, 2004.

References

[261] S. Schlögl, W. M. Heckl, and M. Lackinger. On-Surface Radical Addition of Triply Iodinated Monomers on Au(111) - The Influence of Monomer Size and Thermal Post-Processing. *Surface Science*, 606(13-14):999–1004, 2012.

[262] M. O. Blunt, J. C. Russell, N. R. Champness, and P. H. Beton. Templating Molecular Adsorption Using a Covalent Organic Framework. *Chemical Communications*, 46(38):7157–7159, 2010.

[263] M. O. Blunt, J. C. Russell, M. D. Gimenez-Lopez, N. Taleb, X. L. Lin, M. Schröder, N. R. Champness, and P. H. Beton. Guest-Induced Growth of a Surface-Based Supramolecular Bilayer. *Nature Chemistry*, 3(1):74–78, 2011.

[264] D. F. McMillen and D. M. Golden. Hydrocarbon Bond Dissociation Energies. *Annual Review of Physical Chemistry*, 33:493–532, 1982.

Acknowledgements

Without the many people who helped and supported me over the last four years, this thesis would not be as it is. First, I want to thank my supervisor PD Dr. Markus Lackinger for giving me the opportunity to work in his research group. Numerous discussions on the fundamental principle of the assembly of 2D networks and about technical issues significantly improved my understanding. A special word of gratitude to Prof. Dr. Wolfgang M. Heckl, who set up the general framework for my PhD. I especially thank Prof. Dr. Stefan Sotier for his tremendous support over the last five years in the scientific field as well as in personal matters.

I am also grateful to my colleagues in the STM research group: Dr. Michael Bauer, Niklas Berninger, Dr. Izabela Cebula, Jürgen Dienstmaier, Johanna Eichhorn, Dr. Rico Gutzler, Christoph Heininger, Dr. Marek Janko, Thomas Sirtl, Wentao Song, Maria Wieland and Dr. Hermann Walch for his collaboration in the UHV projects and his suggestions for improvement within this thesis. It is a pleasure to thank Stephan Kloft as well as Stefan Schlögl for invaluable help in any subject, their time for numerous discussions on various subjects, and their friendship. Dr. Alexander M. Gigler and Dr. Marc Hennemeyer, who also have become very close friend to me, have contributed immensely to my personal and professional life and given their wise advice in many areas.

Representative for the machine shop I would also like to acknowledge Günter Hesberg for manufacturing the parts for the UHV system. Moreover, I'd like to thank the technical staff from the department and the electronics workshop who supported this work in numerous ways. In addition, I would like to express my appreciation to my collaboration partners from the Group of Prof. Schmittel (University of Siegen) and Dr. Natalia Martsinovich (University of Warwick).

Acknowledgements

I am also grateful for funding from the Deutsche Forschungsgemeinschaft (DFG), the Center for NanoScience (CeNS), and the Nanosystems Initiative Munich cluster of excellence (NIM), who set up the general financial framework. Personally, I would like to acknowledge the Hanns-Seidel-Stiftung, in particular Dr. Rudolf Pfeifenrath, for monetary and non-material support during my undergraduate studies and my PhD thesis as well as the Kurt Fordan Förderverein für herausragende Begabung e.V. for funding my research stay at the University of Nottingham. I would like to express my gratitude to everybody, who contributed in making my stay in Nottingham a pleasure, in particular to my roommate Sonali Warriar. Special thanks go to Prof. Peter H. Beton for his hospitality, his trust, and many revealing discussions.

My deepest thanks go to my many friends and groups that became a part of my life during the last years for all the emotional support, comradeship, entertainment, and caring they provided, particularly to Barbara Bochter for proof-reading.

I owe my deepest gratitude to my parents-in-law Hannes and Christine and siblings-in-law Katharina and Andreas as well as my parents Georg and Maria and my brothers Christian and Martin for their unconditional support. Lastly, I am eternally grateful to the most important person in my life, my wife Susanne. She has always supported me over the last nine years in all areas of life and always believed in me. Without her I would be a very different person today.

yes
i want morebooks!

Buy your books fast and straightforward online - at one of world's fastest growing online book stores! Environmentally sound due to Print-on-Demand technologies.

Buy your books online at
www.get-morebooks.com

Kaufen Sie Ihre Bücher schnell und unkompliziert online – auf einer der am schnellsten wachsenden Buchhandelsplattformen weltweit! Dank Print-On-Demand umwelt- und ressourcenschonend produziert.

Bücher schneller online kaufen
www.morebooks.de

VDM Verlagsservicegesellschaft mbH
Heinrich-Böcking-Str. 6-8   Telefon: +49 681 3720 174   info@vdm-vsg.de
D - 66121 Saarbrücken   Telefax: +49 681 3720 1749   www.vdm-vsg.de

Printed by Books on Demand GmbH, Norderstedt / Germany